PROMPT ENGINEERING
USING CHATGPT

PROMPT ENGINEERING USING CHATGPT

Crafting Effective Interactions and Building GPT Apps

MEHRZAD TABATABAIAN, PHD, PENG

MERCURY LEARNING AND INFORMATION
Boston, Massachusetts

Publisher: David Pallai
MERCURY LEARNING AND INFORMATION
121 High Street, 3rd Floor
Boston, MA 02110
info@merclearning.com
www.merclearning.com
800-232-0223

M. Tabatabaian. *Prompt Engineering Using ChatGPT: Crafting Effective Interactions and Building GPT Apps.*
ISBN: 978-1-50152-241-3

Library of Congress Control Number: 2024932238

242526321   This book is printed on acid-free paper in the United States of America.

This work is respectfully dedicated to the visionaries advancing our technological frontiers and to each reader endeavoring to use AI for the betterment of humanity.

CONTENTS

PREFACE

Preface

Welcome to an expedition into the realm of prompt engineering, where the art of language meets the science of artificial intelligence (AI). This book is a gateway to understanding the intricate process of crafting effective prompts, with a special emphasis on ChatGPT and its advanced iterations, including GPT-4 and GPTs, the ground-breaking innovations from OpenAI (*https://openai.com/*). Whether you are a student fascinated by the vast potential of AI or a professional trying to harness its power for practical applications, this guide is designed to provide you with the essential knowledge and techniques for effective interaction with AI systems.

AI for All

Our journey is for a diverse audience, encompassing everyone from AI novices to seasoned practitioners. The world of GPT-based models can be complex and intimidating, especially for those without a background in data or computer science. Our mission is to make this world accessible to all. Through a blend of clear, natural language and practical, real-world examples, we aim to demystify the nuances of prompt engineering, equipping you with the tools and confidence to communicate effectively with AI systems.

A Deep Dive into ChatGPT and Beyond

Although there are so many AI innovations, like Gemini™, Bing Chat™, Claude™, and Llamas™, our focus is on ChatGPT and its iterations. This choice was driven by our goal to introduce you to the foundational principles of prompt engineering and empower you to fully leverage the capabilities of ChatGPT and similar advanced models. We want you to understand all the possibilities within the field of AI, relevant to prompt engineering.

The Ever-Evolving Field of AI

The field of AI is dynamic and ever-changing, and so is the art and science of prompt engineering. As you read this book, you will explore the foundational techniques, delve into practical strategies, and discover real-world applications of prompt engineering. In an age where AI technology continues to rapidly change, mastering the skill of crafting effective prompts is invaluable. It paves the way for creative, efficient, and ethical interactions with AI systems, unlocking new opportunities in various domains. This rapid advancement in AI brings with it a host of challenges, from managing the effects of the speed of AI development to harnessing the power of AI/GPT for AI. This book aims to equip you with the knowledge and skills needed to navigate this evolving landscape successfully.

From Theory to Practice: Real-World Insights

This book is not just a theoretical exploration; it is a practical guide. We directly apply the principles of prompt engineering using ChatGPT-3.5 and GPT-4. Through these applications, we demonstrate the creation of prompts that lead to clear, effective communication. The content generated using these models has been thoroughly reviewed and refined for clarity and effectiveness, throughout this book's content. However, as AI technology evolves, so do its responses. The examples in this book serve as a snapshot of current capabilities, with the understanding that AI's language and responses will continue to advance.

A Personal Invitation to AI Mastery

We thank you for choosing to embark on this journey of discovery with us. As we guide you through the dynamic and ever-evolving landscape of AI-generated content, we share our insights, experiences, and passion for one of the most transformative aspects of modern technology. This book is a guide to help you explore AI, a tool to unlock the mysteries of prompt engineering, and a bridge to the future of human-AI interaction.

Managing the Ethics involved with AI and GPT

As we embark on this exploration of prompt engineering, it is imperative to navigate the intricate ethical landscape intertwined with the swift advancement of technologies like GPT and AI. The rapid progress of GPT and AI raises an array of pressing questions and considerations. These encompass ethical dilemmas, moral implications, the establishment of regulatory frameworks, the need to address dominance limitations, issues pertaining to social justice, the necessity of acquiring new skills, the disruptive impact on job markets, and the preservation of fundamental human rights—all serious issues that require prompt attention. As we explore the dynamic realm of prompt engineering, it is vital to acknowledge that these concerns are not mere afterthoughts but integral facets of the AI journey. They serve as a clarion call, summoning the imperative for ongoing research and development within the field of AI.

This book encompasses the following chapters:

Chapter 1: Foundations of Prompts. This chapter provides the fundamental groundwork for understanding prompts, their role in guiding AI models, and their impact on the quality of generated responses.

Chapter 2: Crafting Contextual Prompts. This chapter explores the art of constructing prompts that allow for rich and context-aware interactions with AI models.

Chapter 3: Asking Specific Questions. This chapter describes techniques for formulating clear and direct questions to obtain precise answers from AI models.

Chapter 4: Providing Constraints and Guidelines. This chapter demonstrates how to set constraints and guidelines in your prompts to steer AI responses towards desired outcomes and ethical considerations.

Chapter 5: Creative Prompts for Diverse Content. This chapter explores the creative possibilities of prompts, including those for generating content, ideas, and multimodal responses.

Chapter 6: Debugging and Iterating Prompts. This chapter discusses strategies for evaluating and improving prompts, ensuring they yield the desired results in interactions with AI models.

Chapter 7: Advanced Prompt Engineering. This chapter investigates advanced techniques, including plugins and conditional logic, to harness the full potential of AI models.

Chapter 8: Effective Use of Prompts and GPT-4 Plugins. This chapter discusses how to leverage plugins to extend the capabilities of GPT-4 and create more versatile AI-powered applications.

Chapter 9: Real-World Applications. This chapter demonstrates how prompt engineering is applied across various domains, from customer support and education to business and technical fields.

Chapter 10: Future Trends and Ethical Considerations. This chapter explores the evolving landscape of AI-generated content and delves into ethical considerations as AI technology continues to advance.

Chapter 11: GPTs and GPT Application Builder. This chapter is designed to guide users through the versatile capabilities of the GPTs framework and GPT Builder tool.

Appendix A: Miscellaneous Topics. This section covers a range of additional topics related to prompt engineering, offering practical insights and resources.

Appendix B: Glossary. This section provides a list of important terms and definitions to aid in understanding the terminology of prompt engineering and AI technology.

Acknowledgment

I wish to extend my gratitude to David Pallai from Mercury Learning and Information, my publisher, for his invaluable support.

Mehrzad Tabatabaian, PhD, PEng
Vancouver, BC
May 2, 2024

INTRODUCTION

In the rapidly evolving landscape of artificial intelligence (AI), ChatGPT has emerged as a powerhouse, offering unprecedented capabilities in generating human-like text and speech. As AI becomes an integral part of various domains, understanding how to effectively craft prompts for ChatGPT is essential to unlock its full potential and achieve desired outcomes.

At the heart of this exploration lies the recognition that well-written prompts can ensure seamless communication with ChatGPT. Drawing a parallel to the role of a skilled pilot skillfully guiding an aircraft through the skies, well-constructed prompts take on the critical task of steering and shaping the trajectory of responses crafted by ChatGPT.

ChatGPT is a well-trained mathematical model at the forefront of conversational AI. At its core, it relies on the next-token mathematical model, a fundamental framework that underpins its remarkable conversational abilities. Operating on the principles of probability and deep learning, this model has been meticulously trained on an extensive and diverse dataset. It excels at comprehending the intricacies of language, allowing it to analyze the context and generate responses with remarkable coherence and relevance. When presented with a user's input, it deploys intricate algorithms to predict the most likely next word or token, resulting in responses that not only feel natural, but are also tailored to the ongoing conversation. This mathematical foundation empowers ChatGPT to excel in a wide array of conversational contexts, providing users with an exceptional conversational experience that is coherent, contextually aware, and highly adaptable.

Prompt engineering is the beginning of the journey into the art and science of shaping interactions with this cutting-edge AI language model. This book is your guide to mastering this orchestration, understanding the nuances of prompts, and harnessing their power for a wide array of applications. As we embark on this journey, we will uncover the intricacies of prompt engineering, delve into the psychology of effective communication, and discover how to make ChatGPT's responses align with specific objectives. There are many possibilities, from refining the phrasing of a question to imbuing context into a conversation.

Adding to this is the revolutionary emergence of GPT App Builders, platforms enabling individuals and organizations to design custom applications powered by the linguistic capabilities of GPT. These builders offer a user-friendly interface to create, experiment, and deploy applications tailored to specific needs or industries. By incorporating GPT App Builders, we can refine the conversation and revolutionize the application of conversational AI in various sectors. They serve as the bridge between technical prowess and practical application, making it possible for anyone, from developers to business and technical leaders, to create personalized AI-driven solutions.

Join us in unraveling the art of prompt engineering, a skill that empowers us to bridge the gap between human thought and AI-generated text, and to create a harmonious partnership that transforms the way we interact with technology.

Figure I.1 shows the relationships among AI, NLP, and other subfields.

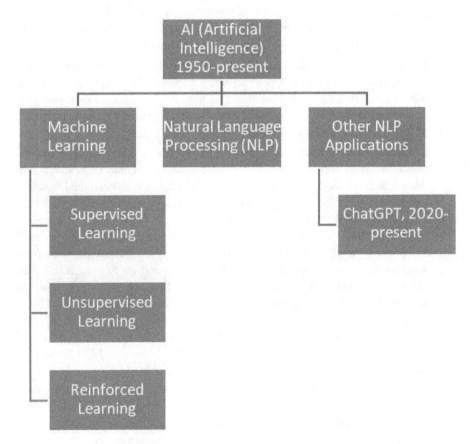

FIGURE I.1 A flowchart showing AI and its subfields.

HISTORY OF LLM AND GPT

Language Models (LMs) and Large Language Models (LLMs) like GPT (Generative Pre-trained Transformer) have revolutionized the field of natural language processing (NLP) and data science. The history of LLMs can be traced back to the early days of machine learning and artificial intelligence, where researchers began exploring methods to teach machines to understand and generate human language. Early attempts, such as rule-based systems and statistical language models, had limitations in capturing the nuances and complexities of human language.

The breakthroughs in LLMs came in the form of deep learning and neural networks. The concept of pre-training, where models learn from vast amounts of text data before fine-tuning on specific tasks, became pivotal. In 2018, OpenAI introduced the first GPT model, GPT-1. This model, with 117 million parameters, showed the potential of pre-trained language models for various NLP tasks. It laid the foundation for subsequent advancements. GPT-3, developed by OpenAI, made headlines in 2020 for its astonishing ability to generate coherent and contextually relevant text. It achieved this by leveraging a massive transformer architecture with 175 billion parameters, allowing it to capture intricate patterns and associations within language. In 2023, OpenAI released GPT-4, which was trained on 1.76 trillion parameters, making it the most advanced system to date.

Since then, the field has continued to evolve. Researchers are exploring ways to make LLMs more efficient, ethical, and applicable across various domains. Additionally, LLMs are being employed in diverse applications, including engineering and manufacturing, healthcare, content generation, education, and natural language understanding in virtual assistants. The list of applications is growing as time goes by, complemented with plugins for GPT-4 and multimodal models.

For the latest information on LLMs and GPT technologies, readers can refer to recent publications in top-tier conferences such as the Conference on Neural Information Processing Systems (NeurIPS) and the International Conference on Machine Learning (ICML). These conferences feature cutting-edge research in the field and provide insights into the latest advancements in LLM technology.

ABOUT THE AUTHOR

Dr. Mehrzad Tabatabaian is a faculty member at the Mechanical Engineering Department, School of Energy at the British Columbia Institute of Technology (BCIT). He has several years of teaching and industry experience. Dr. Tabatabaian is currently Chair of the BCIT School of Energy Research Committee. He has published several papers in scientific journals and for conferences and has written textbooks on Multiphysics and turbulent flow modelling, advanced thermodynamics, tensor analysis, direct energy conversion, the Bond Graph modelling method, and calculus. He holds several registered patents in the energy field, the results of years of research activities.

Dr. Tabatabaian volunteered to help establish the Energy Efficiency and Renewable Energy Division (EERED), a new division at Engineers and Geoscientists British Columbia (EGBC).

Dr. Tabatabaian received his BEng from Sharif University of Technology (formerly AUT) and holds advanced degrees from McGill University (MEng and PhD). He has been an active academic, professor, and engineer in leading alternative energy, oil, and gas industries. Dr. Tabatabaian also has a Leadership Certificate from the University of Alberta and holds an EGBC P.Eng. License.

1

FOUNDATIONS OF PROMPTS

Prompts are the guides that reveal the potential of ChatGPT/GPT, shaping its responses and steering conversations. This chapter provides an introduction to effective prompt engineering, delving into the fundamental principles that underpin the art of crafting prompts to elicit the desired outputs.

1.1 THE ROLE OF PROMPTS IN INTERACTING WITH CHATGPT

Prompts enhance AI-generated interactions, influencing the nature and quality of the responses. Understanding prompts as the input mechanism for ChatGPT is essential. We explore the intricacies of how prompts influence model behavior, emphasizing their significance in initiating meaningful dialogues.

1.2 ANATOMY OF A WELL-CONSTRUCTED PROMPT

Effective prompts are crafted with precision, and their structure plays a pivotal role in yielding accurate and relevant responses. We delve into the components of a well-constructed prompt, from the initial query to the inclusion of context and any specific guidelines. Understanding these elements enhances the efficacy of prompt engineering.

The Essence of Clarity and Specificity

A well-constructed prompt must possess clarity and specificity. There should be no ambiguity in the text, guiding the LLM toward the desired outcome. Consider this example: "Tell me about engineering." This vague prompt is like sending an LLM on a journey without a map; the result is a rambling, unstructured response. Contrast this with a well-crafted prompt like "Provide an overview of the mechanical engineering discipline." This specific prompt narrows the LLM's focus, enabling it to produce a concise and informative response.

Harnessing the Power of Context

Context is very important in prompt engineering. A well-constructed prompt should leverage context to its advantage. For instance, compare the prompts "Explain the theory of relativity" and "Given the context of physics, explain the theory of relativity." The latter, by explicitly providing context, guides the LLM to generate a response that aligns with the field of physics, leading to a more relevant and accurate answer.

Avoiding Leading or Biased Language

A pitfall to avoid in prompt engineering is the use of leading or biased language. Effective prompts maintain neutrality to obtain unbiased responses. For example, consider the prompts "Why are renewable energy sources superior to fossil fuels?" and "Compare the advantages and disadvantages of renewable energy sources and fossil fuels." The first prompt subtly suggests a bias towards renewable energy, while the second one maintains a neutral stance, prompting a balanced response.

In essence, the anatomy of a well-constructed prompt hinges on clarity, specificity, context, and neutrality. By mastering these principles, we empower ourselves to navigate the vast landscape of possibilities that LLMs offer, ultimately elevating our interactions and enhancing our ability to extract meaningful insights.

1.3 EXPLORING DIFFERENT PROMPT STYLES

Prompt engineering is an art that goes beyond mere syntax; it encompasses style, tone, and structure. The choice of prompt style can significantly impact the outcome of interactions with LLMs such as ChatGPT3.5 and GPT-4. In this section, we delve into diverse prompt styles, each with its unique strengths and applications. The practical examples of prompts in this chapter were developed using insights from recent applications of ChatGPT.

Interrogative Prompts: Seeking Direct Answers

Interrogative prompts are question-based and are highly effective when you seek precise, direct responses. They frame the input as a query, prompting the LLM to answer concisely. For example, "What are the main components of a cell?" or "Can you explain the concept of blockchain technology?" Interrogative prompts establish a clear expectation for informative, to-the-point responses, making them valuable in educational contexts and information retrieval tasks.

Imperative Prompts: Issuing Commands

Imperative prompts are directive and instructive. They are used when you want the LLM to perform a specific action or generate content according to a set of guidelines. For instance, "Summarize the key findings of this research paper" or "Write a creative story about a time traveler." Imperative prompts empower the user with control, directing the LLM's output toward a predefined objective. These prompts are instrumental in content generation tasks and automating repetitive writing tasks.

Declarative Prompts: Providing Context or Information

Declarative prompts are informative and provide context or background information. They can be especially useful when you want to guide the LLM with a specific context. For example, "In the context of environmental science, discuss the impact of deforestation" or "Given the history of the Roman Empire, analyze its fall." Declarative prompts help the LLM understand the context and generate content that aligns with it, facilitating more informed and context-aware responses.

Conversational Prompts: Fostering Dialogue

Conversational prompts are designed to initiate and sustain a natural, interactive conversation with the LLM. They often include cues like "Hello" or "Tell me more about" to encourage the model to engage in a back-and-forth dialogue. For example, "Hello, can you help me understand quantum physics?" or "Tell me more about your thoughts on climate change." Conversational prompts are key to creating engaging chatbots and virtual assistants that mimic human conversation.

In conclusion, understanding different prompt styles is essential for effectively engaging with LLMs across various applications. The choice of prompt style should align with your goals and the nature of the task. By mastering the art of prompt style selection, you can unlock the full potential of prompt engineering and enhance the quality of interactions with LLMs.

1.4　PROMPT EXAMPLES AND ANALYSIS

Crafting effective prompts is an art that requires precision and creativity. In this section, we explore a range of prompt examples, each tailored to specific domains, including engineering, to highlight the nuances and strategies behind their construction. These examples are grounded in the latest research and practical insights to illustrate the diversity of prompt styles and applications.

Engineering and Technical Fields

EXAMPLE 1: *"Explain the principles of aerodynamics governing the lift and drag forces in the context of aircraft design."*

Analysis: This prompt exemplifies the use of declarative style in the context of technical fields. It provides specific context ("in the context of aircraft design") and a clear directive, asking the LLM to explain the principles of aerodynamics related to lift and drag forces. Such prompts are instrumental in obtaining structured and detailed responses, making them invaluable in engineering research and education.

Creative Writing and Content Generation

EXAMPLE 2: *"Compose a 500-word article on the future of sustainable architecture, emphasizing innovative materials and energy-efficient design."*

Analysis: Here, an imperative prompt style is employed to direct the LLM to create content with specific guidelines. By specifying the word count and emphasizing key aspects, such as "innovative materials" and "energy-efficient design," the prompt ensures that the generated article is tailored to the topic of sustainable architecture and serves as a foundation for further editing or publication.

Healthcare and Medical Research:

EXAMPLE 3: *"Discuss the recent advancements in gene-editing technologies like CRISPR-Cas9, highlighting their potential applications in treating genetic disorders."*

Analysis: This interrogative prompt solicits a comprehensive response by asking the LLM to "discuss" and "highlight." It specifies the focus on "recent advancements" and "potential applications," indicating the need for an up-to-date, informative response. In the domain of healthcare and medical research, precision and currency are paramount, and this prompt style reflects that.

Conversational AI and Virtual Assistants:

EXAMPLE 4: *"Hello, can you tell me about the weather forecast for this weekend in Vancouver City?"*

Analysis: Conversational prompts, as seen in this example, often start with a friendly greeting to establish a natural conversation flow. By asking for the weather forecast in a specific location and timeframe, the prompt encourages the LLM to provide context-aware information, mimicking the interaction with a virtual assistant or chatbot.

Educational and Learning Applications:

EXAMPLE 5: *"Explain the fundamental concepts of quantum mechanics in a way that a high school student can understand."*

Analysis: This imperative prompt combines a directive style with a specific audience focus ("a high school student"). It challenges the LLM to simplify complex subject matter for educational purposes, showcasing the adaptability of prompt engineering for knowledge dissemination and pedagogical applications.

Role-Goal-Context Style:

One of the most versatile and powerful styles is the Role-Goal-Context (RGC) prompt style. RGC prompts excel in providing specific and context-aware interactions with AI models. By explicitly defining three key components (the "Role," the "Goal," and the "Context"), this prompt style can be harnessed for a wide range of applications.

The RGC structure is as follows:

- **Role**: This component identifies the persona or function you want the AI model to adopt when generating responses. It sets the stage for the AI to embody a particular character, profession, or expertise. For instance, it could be a scientist, a teacher, or a customer support representative.
- **Goal**: The goal articulates the specific objective you want to achieve through the AI's response. It defines the task, the problem to solve, or the information to provide. It is the "what" that you want the AI to do.
- **Context**: Context supplies the necessary background information or constraints that influence the AI's response. It can encompass situational details, user preferences, or any pertinent information required for the AI to generate a relevant and meaningful answer.

The following prompt example refers to a customer support case.

EXAMPLE 6: *"As a customer support agent, provide troubleshooting for connectivity issues for a user experiencing frequent disconnections using a Wi-Fi router."*

Analysis: In customer support, this style ensures that the AI responds with technical troubleshooting guidance, understanding the user's context and issue. It has the following format:

1. *Role*: Customer support agent

2. *Goal*: Troubleshoot connectivity issues for a user experiencing frequent disconnections.

3. *Context*: User is using a Wi-Fi router.

CRAFTING CONTEXTUAL PROMPTS

For AI-powered communication, crafting contextual prompts is akin to composing a well-crafted question or preparing an engaging conversation. Contextual prompts lay the foundation for eliciting coherent and relevant responses from ChatGPT. In this chapter, we delve into the art of designing prompts that effectively guide the model to provide nuanced and targeted replies.

2.1 LEVERAGING CONTEXT FOR MORE RELEVANT RESPONSES

Effective communication hinges on understanding context, and ChatGPT is no exception. By infusing prompts with context, we guide the model to better comprehend the conversation's trajectory. Whether it is referring to prior messages in a chat or incorporating domain-specific cues, contextual prompts empower ChatGPT to respond with coherence and relevance. Context encompasses the information and background that surrounds a conversation or a task. Effective prompt engineering involves framing questions or requests in a way that provides context to the LLM, aiding its understanding and response generation. This is especially crucial for open-domain conversational AI, where context can rapidly evolve within a conversation.

EXAMPLE 1: Basic Context Enhancement

Prompt: "In the context of space exploration, explain the concept of black holes."

Analysis: In this prompt, the inclusion of "in the context of space exploration" provides vital context to the LLM. It specifies that the discussion should revolve around black holes within the framework of space exploration. Without this context, the response could be less focused or might lack relevance to the specified domain. Such contextual cues are fundamental in ensuring that the LLM generates responses that align with the desired topic and domain.

EXAMPLE 2: Utilizing Previous Messages

Prompt: "What are the key ingredients for a classic Margherita pizza?"

Previous Message:

User: "I love Italian cuisine, especially pizza."

Analysis: In this scenario, the user's previous message provides contextual information about their culinary preferences. The prompt leverages this context to tailor the response, focusing on Italian cuisine and, specifically, Margherita pizza. This results in a more contextually relevant answer that aligns with the user's stated preference.

EXAMPLE 3: Dynamic Contextual Prompt

Prompt: "Tell me more about the history of space exploration."

Additional User Input: *"I'm particularly interested in the Apollo missions."*

Analysis: This dynamic approach adapts prompts based on evolving context within a conversation. The prompt adjusts based on the user's input, becoming more specific by incorporating the user's interest in the Apollo missions. This dynamic contextual adaptation ensures that the LLM's response is tailored to the user's current focus within the broader topic of space exploration.

In conclusion, the art of leveraging context for more relevant responses is a multifaceted aspect of prompt engineering. Whether through framing prompts with domain-specific context, utilizing conversational history, or dynamically adapting to real-time user input, context serves as a powerful tool in enhancing the quality and relevance of interactions with LLMs like ChatGPT.

2.2 HARNESSING PRIOR CHAT TURNS FOR SMOOTH CONVERSATIONS

One of the defining characteristics of modern conversational AI models, such as ChatGPT and its variants, is their ability to maintain context and reference previous chat turns. This feature facilitates seamless and coherent conversations, closely mirroring human interactions. In this section, we explore the art of using previous chat turns to enhance conversational continuity and depth, supported by both research findings and practical examples.

EXAMPLE 1: Contextual Continuity

Prompt: "Tell me about the historical significance of the Great Wall of China."

Previous Chat Turn:

User: "I'm planning a trip to China next summer."

Analysis: The user's previous message about planning a trip to China provides context for the prompt. By referencing this context, the conversational AI can tailor the response to focus on aspects of the Great Wall of China that may be relevant to a traveler, such as its historical and tourist significance. This demonstrates how continuity enhances relevance within a conversation.

EXAMPLE 2: Multi-Turn Context

Prompt: "Explain the process of photosynthesis in plants."

Previous Chat Turns:

User: *"What is photosynthesis?"*

AI: "Photosynthesis is the process by which plants convert sunlight into energy."

Analysis: In this example, the user's initial question is followed by the AI's explanation. The user then asks for more information, providing an opportunity for the AI to build upon the previous explanation. This multi-turn context enables the AI to provide a comprehensive response, breaking down the complex process of photosynthesis step by step.

EXAMPLE 3: Handling Complex Dialogue

Prompt: "Debate the pros and cons of renewable energy sources versus fossil fuels."

Previous Chat Turns:

User: *"What are the environmental benefits of renewable energy?"*

AI: "Renewable energy sources, like wind and solar power, produce fewer greenhouse gas emissions."

Analysis: In this scenario, the user initiates a discussion about renewable energy's environmental benefits. The AI's response can refer to its previous statement about reduced greenhouse gas emissions to begin a thoughtful debate on the broader topic of renewable energy versus fossil fuels.

In conclusion, using previous chat turns for seamless conversations is a cornerstone of effective conversational AI. By retaining and referencing context, these models are equipped to engage in deeper, more relevant, and more coherent dialogues. This capability is instrumental in creating AI systems that are not only informative but also capable of sustaining engaging and contextually rich interactions.

2.3   INCORPORATING THE USER'S NAME AND DETAILS FOR PERSONALIZATION

Personalization in conversational AI enhances engagement and relevance. By using details like a user's name and preferences, AI models can forge a stronger connection and deliver responses that are more customized to the individual. In this section, we explore the strategies and benefits of incorporating user-specific information into conversations, with supporting references and practical examples. One way of doing this is with the "Custom instructions" available in the user's settings (*https://help.openai.com/en/articles/8096356-custom-instructions-for-chatgpt*).

There are two spaces available for providing user's info and the way that the user wants ChatGPT to respond, as shown in Figure 2.1.

Custom instructions ⓘ

What would you like ChatGPT to know about you to provide better responses?

0/1500   Hide tips ⌄

How would you like ChatGPT to respond?

0/1500

Enable for new chats ⬤   Cancel   Save

FIGURE 2.1 Custom instructions window in ChatGPT.

EXAMPLE 1: Simple Personalization

Prompt: "Hello, my name is [User's Name]. how can you assist me with [the topic] today?"

Analysis: By incorporating the user's name into the greeting, the AI immediately establishes a personal connection with the user. This simple act of recognition can create a positive and welcoming atmosphere, enhancing the user's engagement and comfort within the conversation.

EXAMPLE 2: User Preferences

Prompt: "I noticed you're interested in [User's Interest]. Would you like to know more about the latest developments in that field?"

Previous Chat Turn:

User: *"I'm really passionate about renewable energy."*

Analysis: In this example, the AI references the user's previously expressed interest in renewable energy. This reference not only demonstrates active listening but also tailors the subsequent conversation to the user's preferences, potentially leading to a more engaging and relevant discussion.

EXAMPLE 3: Dynamic Personalization

Prompt (Adapting to User Input): "Tell me more about your recent trip to [User's Mentioned Location]. How was your experience there?"

User Input:

User: *"I just returned from a trip to Paris."*

Analysis: The prompt dynamically incorporates the user's mentioned location (Paris) and their recent travel experience. This contextual personalization not only demonstrates attentiveness but also enables the AI to engage in a more meaningful conversation about the user's trip.

EXAMPLE 4: Demonstrating Empathy

Prompt: "I understand that you're a parent, [User's Name]. How can I assist you in finding family-friendly activities in your area?"

Analysis: In this example, the AI not only uses the user's name but also acknowledges their role as a parent. This empathetic approach can help build trust and rapport, as the user feels understood and supported in their specific context.

In conclusion, incorporating user-specific details for personalization is a potent strategy in creating engaging and relevant AI-driven conversations. Whether by using the user's name, referencing their preferences, or adapting to the conversation's evolving context, personalization enhances the user experience and contributes to more meaningful interactions.

CHAPTER

3

Asking Specific Questions

In the realm of AI-driven interactions, the ability to ask clear and specific questions through prompts becomes a powerful skill. Chapter 3 explores the art of crafting prompts that not only evoke precise responses but also navigate the complexities of language to extract accurate information from ChatGPT.

3.1 TECHNIQUES FOR ASKING CLEAR AND DIRECT QUESTIONS

The art of asking clear and direct questions is fundamental to effective communication, especially when engaging with AI systems. In this section, we will explore various techniques and strategies for formulating precise questions that yield accurate and relevant responses. These techniques are informed by research findings and are accompanied by practical examples to illustrate their application.

EXAMPLE 1: The Power of Precision

Prompt: "Please explain the process of cellular respiration in eukaryotic cells."

Analysis: This prompt exemplifies the technique of precision. It specifies the topic (cellular respiration) and the context (eukaryotic cells), leaving no room for misinterpretation. By eliminating ambiguity, the AI can provide a concise and accurate response, catering to the user's request.

EXAMPLE 2: Clarity Through Context

Prompt: "In the context of economics, what are the key principles of supply and demand?"

Analysis: This prompt leverages context to enhance clarity. By framing the question within the context of economics, the AI understands the user's area of interest and provides a response that pertains specifically to supply and demand principles within that domain. Contextual framing reduces ambiguity and ensures relevance.

EXAMPLE 3: Closed-Ended Clarity

Prompt: "Is the melting point of gold higher than that of iron?"

Analysis: By structuring the question as a closed-ended query, the user seeks a straightforward response that compares the melting points of gold and iron. This format eliminates the need for lengthy explanations and encourages the AI to provide a concise, fact-based answer.

EXAMPLE 4: Ambiguity Clarified

Prompt: "Can you tell me about the recent advancements in AI and machine learning?"

Analysis: This prompt initially raises a potential issue of ambiguity. "Recent advancements" could refer to developments in the last year, last month, or even more recently. To eliminate this ambiguity, the user could specify a time frame, such as "Can you tell me about advancements in AI and machine learning in the past year?" This modification ensures a more precise response.

In conclusion, asking clear and direct questions is a foundational skill when interacting with AI systems. Employing techniques such as specificity, contextual framing, closed-ended queries, and ambiguity avoidance can significantly enhance the quality of responses received, ultimately leading to more productive and meaningful interactions.

3.2 NAVIGATING AMBIGUITY: HOW TO GET PRECISE ANSWERS

Ambiguity is a common challenge when interacting with AI systems, but it can be effectively managed with the right techniques. In this section, we will explore strategies for navigating ambiguity to ensure that you receive precise and relevant answers from AI models. These strategies are grounded in research and are illustrated with practical examples to demonstrate their effectiveness.

The Ambiguity Conundrum

Ambiguity arises when a question or request lacks clarity or specificity. AI models, while powerful, may struggle to provide precise responses when faced with ambiguous queries. To overcome this challenge, it is essential to employ techniques that reduce ambiguity.

EXAMPLE 1: Dealing with Vagueness

Prompt: "Tell me about the best restaurants."

Analysis: This prompt is inherently ambiguous because it doesn't specify criteria for what constitutes "best" or a location. To obtain precise answers, it's crucial to provide more context or criteria. For instance, "Can you recommend the top-rated Italian restaurants in New York City?" narrows the scope and reduces ambiguity.

EXAMPLE 2: Clarifying Scope

Prompt: "Explain climate change."

Analysis: The term "climate change" can encompass various aspects, such as causes, impacts, or mitigation strategies. To get a precise answer, specify the aspect you want to know more about. For example, "Can you describe the human causes of climate change?" narrows the focus and reduces ambiguity.

EXAMPLE 3: Multi-Step Clarity

Prompt (Multi-Step): "First, explain the concept of quantum entanglement. Then, describe its implications for quantum computing."

Analysis: When dealing with complex topics like quantum physics, breaking down the question into two distinct steps reduces ambiguity. It ensures that each part of the question receives a focused and precise response.

EXAMPLE 4: Request for Clarification

User: "Tell me about recent advancements in AI."

AI: "AI has seen significant progress recently."

User (Seeking Clarification): "Could you provide specific examples of these advancements?"

Analysis: When faced with a vague response, the user seeks clarification by requesting specific examples. This proactive approach guides the conversation toward a more precise and informative answer.

EXAMPLE 5: Defining Key Terms

Prompt: "Explain the role of 'dark matter' in astrophysics."

Analysis: "Dark matter" is a term with a specific meaning in astrophysics. By enclosing it in quotes or providing a brief definition within the prompt, you reduce the risk of ambiguity and ensure the AI understands your question accurately.

In conclusion, navigating ambiguity to obtain precise answers from AI systems is essential for productive interactions. Employing techniques such as providing context and specifics, breaking down questions into multiple steps, seeking clarification, and defining key terms can significantly enhance the quality and accuracy of the responses you receive.

3.3 UNCOVERING HIDDEN INFORMATION WITH WELL-FORMED QUERIES

In the realm of conversational AI and information retrieval, the ability to uncover hidden or less apparent information is a valuable skill. Well-formed queries, structured to dig deeper or explore nuanced topics, can be the key to extracting valuable insights. In this section, we delve into the techniques and strategies for crafting well-formed queries to unearth concealed information.

EXAMPLE 1: Probing for Details

Prompt: "Tell me about the history of the Eiffel Tower construction, including any lesser-known facts."

Analysis: This prompt goes beyond a generic request for information and specifically instructs the AI to uncover lesser-known details about the Eiffel Tower's construction history. By encouraging the exploration of hidden facts, the AI can provide a richer and more informative response.

EXAMPLE 2: Open-Ended Inquiry

Prompt: "What can you tell me about the cultural impact of the Beatles beyond their music?"

Analysis: By asking about the cultural impact "beyond their music," the query prompts the AI to delve into less obvious aspects of the Beatles' influence, such as their role in shaping fashion, social movements, or popular culture.

EXAMPLE 3: Comparative Exploration

Prompt: "Compare and contrast the approaches of two prominent philosophers, Immanuel Kant and Friedrich Nietzsche, in their views on ethics."

Analysis: This query compels the AI to explore nuanced differences and hidden insights within the ethical philosophies of Kant and Nietzsche. By highlighting the need for both comparison and contrast, it encourages the revelation of less evident distinctions.

EXAMPLE 4: Request for Evidence

Prompt: "Provide examples of real-world applications of blockchain technology in industries beyond finance."

Analysis: By asking for specific examples "beyond finance," this query encourages the AI to uncover less obvious applications of blockchain technology, such as supply chain management, healthcare, or voting systems.

EXAMPLE 5: Historical Contextualization

Prompt: "In the context of the 1960s Civil Rights Movement, discuss the contributions of lesser-known activists and their impact on social change."

Analysis: This query, framed within the historical context of the 1960s Civil Rights Movement, prompts the AI to uncover the often-overlooked contributions of lesser-known activists, shedding light on hidden facets of the movement.

In conclusion, uncovering hidden information with well-formed queries is a powerful tool in the arsenal of conversational AI users. Techniques such as employing open-ended queries, encouraging comparative analysis, requesting supporting evidence, and using historical or contextual prompts can unearth concealed knowledge and provide a deeper understanding of complex topics.

4

PROVIDING CONSTRAINTS AND GUIDELINES

Chapter 4 examines the art of using ChatGPT by setting constraints and guidelines within prompts. By establishing boundaries, we shape the content generated and ensure responsible, ethical, and contextually appropriate responses.

4.1 SETTING CONSTRAINTS FOR DESIRED OUTPUT

In the dynamic landscape of AI and natural language processing, setting constraints on the desired output is a strategic approach for fine-tuning responses and ensuring they align with specific objectives. This section delves into the nuances of constraint setting as a technique for refining AI-generated content. It is underpinned by research findings and enriched with practical examples to elucidate its applications.

The Role of Constraints in AI Interaction

Constraints act as guiding principles that direct AI models to generate responses within defined boundaries. They are instrumental in tailoring responses to meet particular criteria, fostering controlled and contextually appropriate interactions.

EXAMPLE 1: Imposing a Length Constraint

Prompt: "Summarize the impact of climate change on polar ice caps in 50 words or less."

Analysis: By stipulating a word limit of 50 words, this constraint ensures that the AI provides a concise summary. This is invaluable for scenarios where brevity is paramount, such as news headlines or social media posts.

EXAMPLE 2: Format Constraint for a Poem

Prompt: "Compose a Khayyam-style poem about the beauty of a sunset."

Analysis: By requesting a Khayyam, a traditional form of Persian poetry with specific syllable, the prompt guides the AI to generate content within the specified poetic format, resulting in a concise and structured response.

EXAMPLE 3: Domain Constraint in Medical Context

Prompt: "Explain the mechanism of action of antibiotics in treating bacterial infections."

Analysis: By referencing the specific drug and medical condition, this constraint confines the AI's response to the domain of medicine, guaranteeing a focused and expert-level explanation.

EXAMPLE 4: Mimicking a Literary Style

Prompt: "Write a short story in the style of Edgar Allan Poe."

Analysis: By invoking the style of Edgar Allan Poe, this constraint directs the AI to craft a story characterized by Poe's distinctive dark and atmospheric narrative style, offering a tailored literary experience.

EXAMPLE 5: Ensuring Ethical Responses

Prompt: "Please share insights regarding the ethical considerations surrounding interactions between robots and humans in clinical and hospital environments."

Analysis: This constraint underscores the importance of offering help and resources while strictly adhering to ethical guidelines in the context of health industry.

EXAMPLE 6: Custom Constraint for Sensitive Topics

Prompt: "Discuss the historical context of [sensitive topic] while avoiding offensive language or bias."

Analysis: Here, a custom constraint is set to address a sensitive topic while ensuring that the response maintains a respectful and unbiased tone.

Constraints play an indispensable role in harnessing the potential of AI models while aligning their output with specific objectives and ethical considerations. Whether guiding content length, controlling domains, mimicking styles, ensuring safety, or implementing custom restrictions, constraints empower users to interact with AI in a controlled and purposeful manner.

4.2   ENSURING ETHICAL AND RESPONSIBLE RESPONSES

Ethical and responsible AI interactions are paramount in the development and deployment of AI systems. This section discusses the critical aspect of ensuring that AI-generated responses align with ethical guidelines and societal values. It is supported by key research findings and real-world examples to illustrate the importance and application of ethical considerations in AI.

Ethics in AI: A Fundamental Imperative

As AI systems become increasingly integrated into various aspects of society, addressing ethical concerns is essential to ensure responsible AI usage. The following examples highlight the significance of ethical considerations in AI interactions:

EXAMPLE 1: Mitigating Bias

Prompt: "Provide an overview of gender differences in leadership styles."

Analysis: Ensuring that the AI's response is unbiased and free from gender stereotypes is crucial. A responsible AI response should present a balanced view, acknowledging individual variation and avoiding reinforcement of gender biases.

EXAMPLE 2: Handling Sensitive Topics

Prompt: "Discuss the bombing of Hiroshima during World War II and its historical significance."

Analysis: When addressing sensitive historical events like World War II, the AI should provide a respectful and historically accurate response.

EXAMPLE 3: Bullying Prevention

Prompt: "Could you provide advice on effectively coping with feelings of being bullied? I'm looking for three helpful strategies."

Analysis: Responding to users in distress requires a high level of ethical responsibility. AI responses should prioritize immediate assistance, empathy, and guidance to seek help from human professionals or crisis helplines.

EXAMPLE 4: Political Neutrality

Prompt: "Explain the advantages of socialism over capitalism, and vice versa."

Analysis: To maintain political neutrality and avoid bias, AI responses should provide an objective analysis of both systems without promoting one over the other. Ethical AI should encourage critical thinking rather than imposing a particular viewpoint.

EXAMPLE 5: Privacy and Data Security

Prompt: "What personal information do you store about users, and how is it used?"

Analysis: Ethical AI systems prioritize transparency regarding data handling. Responses should explain data storage and usage practices clearly, while also respecting privacy regulations and user consent.

In conclusion, ensuring ethical and responsible responses from AI systems is not merely a best practice; it is an ethical imperative. Mitigating bias, handling sensitive topics with care, responding to crises responsibly, maintaining political neutrality, and respecting privacy and data security are essential aspects of ethical AI interactions. These considerations must be integrated into the development and deployment of AI systems to promote responsible and ethical AI usage.

4.3 COMBINING CONSTRAINTS FOR TAILORED CONTENT

In the pursuit of precise and contextually relevant AI-generated content, the strategic combination of multiple constraints can be a potent approach. This section discusses the art of combining constraints to tailor AI responses, drawing from research insights and practical examples to illuminate the effectiveness of this technique.

The Synergy of Multiple Constraints

The application of multiple constraints allows for a fine-tuned approach to AI content generation. By carefully combining constraints, users can shape responses to meet specific requirements while ensuring ethical, accurate, and contextually appropriate interactions.

EXAMPLE 1: Crafting Ethical Marketing Content

Prompt: "Generate a marketing tagline for our new product, ensuring it is catchy but free from any offensive language or stereotypes."

Analysis: This prompt combines constraints related to marketing effectiveness and ethical considerations. The resulting tagline should be both engaging and socially responsible, reflecting a balance between marketing goals and ethical standards.

EXAMPLE 2: Generating Technical Documentation

Prompt: "Provide technical documentation for our software product in a concise and jargon-free manner."

Analysis: This prompt blends constraints of conciseness and plain language use. It directs the AI to generate documentation that is both user-friendly and easily understandable, catering to a broader audience.

EXAMPLE 3: Crafting Sensitive Medical Content

Prompt: "Explain the side effects of this medication, keeping the language simple and avoiding any alarming language."

Analysis: Here, constraints related to simplicity and avoiding alarmist language are employed. The AI response must ensure clarity while not raising undue concerns among users about potential side effects.

EXAMPLE 4: Content for Education

Prompt: "Write an educational article on climate change for middle school students, using language appropriate for their age and maintaining scientific accuracy."

Analysis: This prompt combines constraints of age-appropriate language and scientific accuracy. The AI's response should strike a balance, making complex scientific concepts accessible to a younger audience while maintaining factual integrity.

EXAMPLE 5: Ethical Sales Pitch

Prompt: "Compose a sales pitch for our eco-friendly product, emphasizing its environmental benefits while avoiding greenwashing or exaggeration."

Analysis: This request merges constraints related to environmental messaging and ethical marketing. The AI-generated sales pitch must promote the product's environmental virtues transparently and honestly.

In summary, combining constraints for tailored content allows users to fine-tune AI responses to meet diverse objectives while upholding ethical, contextual, and user-centric considerations. The synergy of constraints can produce content that strikes a harmonious balance between competing requirements, resulting in responses that are both precise and contextually relevant.

5

CREATIVE PROMPTS FOR DIVERSE CONTENT

Chapter 5 investigates creative prompt engineering, where the fusion of imagination and AI-generated content is important. By mastering the art of crafting open-ended prompts, we unlock ChatGPT's potential to generate a wide range of creative outputs, from stories and poetry to innovative ideas.

5.1 INSPIRING CREATIVE WRITING WITH OPEN-ENDED PROMPTS

Open-ended prompts are a catalyst for creativity in AI-driven content generation. This section explores the art of inspiring creative writing through open-ended prompts. The practical examples of prompts in this chapter were developed using ChatGPT application.

Fostering Creativity Through Openness

Open-ended prompts are characterized by their inherent lack of constraints. They provide the creative freedom necessary for AI models to generate content that is imaginative, innovative, and tailored to specific creative objectives.

EXAMPLE 1: Crafting a Short Story

Prompt: "Write a short story that begins with the sentence: 'The old, dusty book sat unopened on the shelf, waiting for someone brave enough to turn its pages.'"

Analysis: This open-ended prompt offers the freedom to explore countless narrative possibilities. It sparks creativity by presenting an intriguing scenario but leaves the plot, characters, and themes entirely to the writer's imagination.

EXAMPLE 2: Creating Poetry

Prompt: "Compose a poem inspired by the concept of 'silence.'"

Analysis: Poetry often thrives in open-ended settings. This prompt invites writers to explore the multifaceted theme of "silence" in their own unique ways, allowing for diverse interpretations and poetic expressions.

EXAMPLE 3: Designing Artwork Descriptions

Prompt: "Write a compelling description for an abstract painting that evokes a sense of wonder."

Analysis: This prompt empowers writers to craft vivid descriptions without prescribing specific details. It fosters creative expression by encouraging writers to evoke emotions and imagery through their words.

EXAMPLE 4: Developing Dialogues

Prompt: "Write a dialogue between two characters meeting for the first time in a bustling city park."

Analysis: Dialogues are excellent vehicles for character development and storytelling. This open-ended prompt sets the stage for character interaction while leaving the nature of their conversation and the direction of the plot open to the writer's imagination.

EXAMPLE 5: Crafting Innovative Solutions

Prompt: "Describe a futuristic invention that revolutionizes transportation."

Analysis: Futuristic scenarios are ideal for open-ended prompts. This prompt encourages writers to envision and articulate imaginative inventions, pushing the boundaries of innovation and creativity.

In conclusion, open-ended prompts are invaluable tools for inspiring creative writing. They offer writers the freedom to explore, imagine, and innovate without the constraints of rigid guidelines. By leveraging ambiguity, themes, emotional depth, character interactions, and forward-thinking concepts, open-ended prompts unlock the full spectrum of creative potential in AI-driven content generation.

5.2   GENERATING POETRY, STORIES, AND DIALOGUES

The versatility of AI in creative content generation is showcased in its ability to craft poetry, stories, and dialogues. This section explores how AI can be used to produce compelling literary works. The practical examples of prompts in this chapter were developed using ChatGPT application.

AI as a Creative Muse

AI models have emerged as a source of inspiration for poets, authors, and playwrights. They can generate poetry, stories, and dialogues that span a wide range of themes and styles, providing a wellspring of creative ideas.

EXAMPLE 1: Poetry Generation

Prompt: "Compose a poem that captures the essence of a serene forest at dawn."

Analysis: AI can weave words into poetic verses, evoking the tranquil beauty of a forest awakening to the first light of day. Poems like these are not just the result of algorithms but a harmonious blend of data-driven creativity.

EXAMPLE 2: Storytelling

Prompt: "Craft a short story about a time traveler who discovers a forgotten civilization in the depths of the Amazon rainforest."

Analysis: AI-driven storytelling unfolds like a literary journey, immersing readers in imaginative tales that blend elements of science fiction, adventure, and exploration. The result is a narrative that sparks curiosity and wonder.

EXAMPLE 3: Dialogue Creation

Prompt: "Write a dialogue between a wise old sage and a curious young apprentice discussing the secrets of the universe."

Analysis: AI-generated dialogues breathe life into characters, enabling them to engage in meaningful conversations. These interactions may explore philosophical concepts, impart wisdom, or offer valuable insights.

EXAMPLE 4: Collaborative Storytelling

Prompt: "As a creative writing assistant, collaborate with me to write a fiction story."

Analysis: Collaborative storytelling with AI is an interactive and creative process. It combines human imagination with AI's ability to generate coherent and contextually relevant content.

EXAMPLE 5: Historical Dialogue Recreation

Prompt: "Create a conversation between two historical figures, Albert Einstein and Marie Curie, discussing their ground-breaking discoveries."

Analysis: AI-driven dialogue recreation can make historical figures realistic, allowing them to engage in hypothetical conversations that shed light on their thoughts, motivations, and interactions. This enriches historical storytelling and education.

In summary, AI's capacity to generate poetry, stories, and dialogues opens up exciting possibilities for creativity. It serves as a muse, offering thematic, narrative, and character-driven content that can inspire writers and readers alike. The examples provided demonstrate how AI can be a valuable tool for creative content generation, fostering new avenues of literary exploration.

5.3   USING PROMPTS TO GENERATE IDEAS AND CONCEPTS

Prompts are powerful tools not only for generating content but also for helping people develop creative ideas and innovative concepts. This section explores how prompts can be employed to stimulate creativity and idea generation. The practical examples of prompts in this chapter were developed using ChatGPT application.

Prompts as Creative Catalysts

Prompts serve as catalysts for inspiration, and are capable of helping people develop creative ideas, guiding individuals toward novel ideas and concepts.

EXAMPLE 1: Designing Futuristic Technology

Prompt: "Envision a future technology that could revolutionize communication. Describe its functions and user experience."

Analysis: This prompt encourages forward-thinking and innovation in technology design. It challenges individuals to conceptualize breakthroughs in communication, blending creativity with practicality.

EXAMPLE 2: Crafting Unique Culinary Recipes

Prompt: "Create a recipe for a dish that combines traditional cuisine with modern cooking techniques."

Analysis: This prompt inspires culinary creativity, merging the old with the new. It encourages chefs to experiment with flavors and techniques, leading to unique culinary creations.

EXAMPLE 3: Developing Business Strategies

Prompt: "Devise a business strategy for a startup that aims to use AI for social good."

Analysis: In the business world, prompts like this stimulate strategic thinking and entrepreneurial innovation. It challenges business minds to integrate AI technology with social impact initiatives.

EXAMPLE 4: Conceptualizing Architectural Designs

Prompt: "Design an eco-friendly residential building that blends with its natural surroundings."

Analysis: This prompt encourages architects to think about sustainable design and harmony with nature. It fosters creativity in environmental architecture and green living spaces.

EXAMPLE 5: Generating Educational Program Ideas

Prompt: "Develop an idea for an educational program that incorporates virtual reality to enhance learning."

Analysis: This prompt invites educators to think about the integration of technology in learning. It challenges them to create immersive and interactive educational experiences using virtual reality.

In conclusion, prompts are versatile tools for idea generation and concept development across various disciplines. By employing problem-solving frameworks, abstract expression, contextual challenges, genre and theme exploration, and scientific inquiry, individuals can leverage prompts to unlock their creative potential and generate innovative ideas and concepts.

5.4   EXPLORING CREATIVITY THROUGH MULTIMODAL PROMPTS

The convergence of multiple modes of expression, such as text, images, and audio, opens new creative possibilities. This section explores how multimodal prompts can be used for innovative and imaginative content creation using research insights and practical examples. Please note that currently, Chat-GPT and GPT-4 are text based. Tools like OpenAI's CLIP (Contrastive Language-Image Pre-training) and DALL-E are examples that combine vision and language to generate a new result. Recently, GPT-4 obtained the ability to utilize voice and speech as prompts. This feature enables users to talk with GPT-4 while the conversation is saved and becomes available when desired. According to OpenAI, this feature allows for an AI that *"…can see, hear, and speak."* In GPT-4, the "see-hear-speak" capability represents a significant advancement in the model's multimodal abilities. While earlier versions primarily processed and generated text, GPT-4 integrates multiple modalities, allowing it to understand and generate content across visual, auditory, and textual domains. This means that GPT-4 can "see" by processing and interpreting images, "hear" by understanding auditory signals, and "speak" by producing text or potentially other output forms. This multimodal approach enhances the model's versatility, enabling more comprehensive interactions and applications. Whether it is describing the contents of an image, transcribing an audio clip, or generating text based on multi-sensory input, the see-hear-speak feature positions GPT-4 at the forefront of AI capabilities.

The Power of Multimodal Creativity

Multimodal prompts combine various forms of media to inspire creativity. These prompts go beyond using only text and encourage individuals to engage with visual, auditory, and textual elements, resulting in the users enjoying more dynamic and enriched creative processes.

EXAMPLE 1: Visual Storytelling

Multimodal Prompt: "Create a short story inspired by this image: [insert image of a mysterious forest at night]. Use both text and imagery to convey the narrative."

Analysis: This multimodal prompt challenges the AI to not only craft compelling narratives but also incorporate the visual elements of the mysterious forest into its stories. The prompt and its results encourage a synergy between textual and visual storytelling, fostering the user's creativity.

EXAMPLE 2: Music and Poetry Fusion

Multimodal Prompt: "Compose a poem inspired by the melody of this music clip: [insert audio clip of a tranquil piano piece]."

Analysis: This prompt asks the AI to synchronize the words of the poem with the mood and rhythm of the provided music clip, creating a unique fusion of auditory and textual results.

EXAMPLE 3: Interactive Visual Art

Multimodal Prompt: "Produce an interactive digital artwork that changes in response to user interactions. Incorporate both visual and auditory elements to create an immersive experience."

Analysis: This multimodal prompt challenges digital artists to create dynamic and interactive artworks that engage users on multiple sensory levels. It encourages the exploration of user-driven creativity.

EXAMPLE 4: Podcast Scriptwriting

Multimodal Prompt: "Write a script for a podcast episode discussing the future of space exploration. Include not only the dialogue but also visual cues for accompanying animations."

Analysis: This prompt combines textual elements with the anticipation of visual and auditory components. Podcast scriptwriters are tasked with crafting engaging narratives that seamlessly integrate with multimedia content, promoting a holistic approach to storytelling.

EXAMPLE 5: Visual Poetry Performance

Multimodal Prompt: "Create a visual poem that transforms into an audiovisual performance. The poem's text should synchronize with visuals and music, immersing the audience in a multisensory experience."

Analysis: This prompt asks the AI to utilize the written word to develop a multisensory performance combining visual aesthetics, audio compositions, and poetic expression. This prompt utilizes the AI's ability to generate content other than that of traditional poetry.

In conclusion, multimodal prompts are potent tools for fostering creativity through the integration of various media forms. By employing cross-media integration, mood-based creation, immersive experiences, cross-medium storytelling, and sensory integration, creators can explore the novel suggestions generated by the AI that are superior to the results yielded from single-mode prompts.

6

DEBUGGING AND ITERATING PROMPTS

Chapter 6 describes the process of debugging and iterating prompts. Just as an artisan refines a masterpiece stroke by stroke, so too must prompt engineers scrutinize and enhance their prompts to achieve optimal results. This chapter discusses the intricacies of identifying, addressing, and learning from the challenges that arise in prompt design.

6.1   INTERPRETING AND ANALYZING MODEL RESPONSES

Understanding and interpreting model responses is a fundamental aspect of prompt engineering, crucial for crafting effective prompts and iteratively improving model performance. This section examines the nuances of interpreting model responses using research insights and practical examples.

Deciphering Model Outputs

When working with large language models (LLMs) like ChatGPT and GPT-4, interpreting model responses goes beyond evaluating grammatical correctness. It involves a multifaceted analysis that considers factors such as relevance, coherence, bias, and alignment with user intent.

EXAMPLE 1: Assessing Relevance

Prompt: "Explain the concept of black holes."

Analysis: In this scenario, a relevant model response should provide a clear and accurate explanation of black holes in astrophysics. Deviations from this topic, such as unrelated anecdotes or off-topic content, indicate a lack of relevance in the model's response.

EXAMPLE 2: Analyzing Coherence

Prompt: "Write a story about a detective solving a mysterious murder case in a small coastal town."

Analysis: Coherence is vital in narrative prompts. A coherent model response should follow a logical storyline with well-connected plot points, characters, and events. Incoherent storytelling disrupts the narrative flow.

EXAMPLE 3: Identifying Bias

Prompt: "Discuss the advantages and disadvantages of a universal healthcare system."

Analysis: Bias detection is crucial when assessing responses to prompts that involve opinion or analysis. Biased content may present a one-sided view or use emotionally charged language, undermining the objectivity of the response.

EXAMPLE 4: Evaluating Completeness

Prompt: "Explain the process of photosynthesis."

Analysis: Completeness involves assessing whether the model response provides a comprehensive explanation of the topic. Incomplete responses may omit critical details or steps, leaving gaps in the explanation.

EXAMPLE 5: User Intent Alignment

Prompt: "Suggest vegetarian recipes for a family dinner party."

Analysis: Alignment with user intent is essential in prompts with specific requests. The model's response should offer suitable vegetarian recipes tailored to a family dinner party, aligning with the user's intent.

In conclusion, interpreting and analyzing model responses is a multifaceted process that involves evaluating relevance, coherence, bias, completeness, and alignment with user intent. By employing these techniques, prompt engineers can gain deeper insights into model behavior, identify areas for improvement, and iteratively refine prompts to achieve more accurate and contextually relevant responses. We can ask ChatGPT to analyze its responses. For example, the prompt in Example 1 can be written as: "Explain the concept of black holes. Analyze your response in terms of relevance with writing a short paragraph and cite three actual references."

6.2 IDENTIFYING MISUNDERSTANDINGS AND ERRORS

One of the key challenges in prompt engineering is the identification and rectification of misunderstandings and errors in model responses. This section delves into the critical process of recognizing and addressing these issues, drawing from research insights and practical examples.

The Challenge of Misunderstandings

Misunderstandings can occur when the model misinterprets the user's intent, leading to responses that are off-topic, inaccurate, or contextually inappropriate. Identifying and rectifying these misunderstandings is crucial for improving the quality of AI-generated content.

EXAMPLE 1: Ambiguity in Prompts

Prompt: "Explain the significance of the Big Bang."

Analysis: This prompt, while clear to humans, contains ambiguity that an AI model might misinterpret. The model could interpret "Big Bang" as a reference to a TV show or a loud noise, leading to a response unrelated to the intended astrophysical concept.

EXAMPLE 2: Handling Multimodal Prompts

Multimodal Prompt: "Describe the scene depicted in this image: [insert image of a busy city street]."

Analysis: Multimodal prompts, which include visual elements, require carefully choosing the correct terminology and images. If the model cannot interpret the image correctly, it might produce an inaccurate description that does not align with the visual content.

EXAMPLE 3: Addressing Technical Misconceptions

Prompt: "Explain the process of nuclear fusion in stars."

Analysis: In technical prompts, misunderstandings can lead to inaccuracies in responses. If the model misconstrues the scientific concept of nuclear fusion, the resulting explanation may contain errors or inaccuracies. Verification may be required with reliable published scientific sources and references.

EXAMPLE 4: User Intent Clarification

Prompt: "Recommend books for improving mental health."

Analysis: User intent can sometimes be unclear, leading to misunderstandings. In this case, the user may seek books on coping with stress, while the model might recommend books on mental health disorders, resulting in a misalignment of intent.

EXAMPLE 5: Contextual Misalignment

Prompt: "Continue the story from where it left off."

Analysis: Prompts that lack specific context may cause the AI model to produce responses that do not align with the user's expectations or the narrative's continuity, resulting in disruptions or incoherent storytelling.

In conclusion, identifying and rectifying misunderstandings and errors in model responses is a crucial step in prompt engineering. By addressing ambiguity, aligning multimodal content, ensuring technical accuracy, clarifying user intent, and maintaining contextual consistency, prompt engineers can enhance the quality and reliability of AI-generated responses.

6.3   STRATEGIES FOR ITERATING AND IMPROVING PROMPTS

The process of prompt engineering does not end with the initial prompt design; it is an ongoing process of refinement and improvement. In this section, we explore effective strategies for iteratively enhancing prompts to achieve better model responses using research insights and practical examples.

The Iterative Prompt Refinement Cycle

Iterating on prompts is akin to a feedback-driven loop that involves continuous testing, evaluation, and adjustment. This cycle is crucial for fine-tuning model behavior over time.

EXAMPLE 1: Addressing Ambiguity

Initial Prompt: "Explain the concept of 'light.'"

Analysis: The initial prompt is overly broad, potentially leading to responses about various meanings of "light," such as the opposite of darkness or electromagnetic radiation. To address this, the prompt can be refined to specify the context, e.g., "Explain the concept of 'light' in the context of physics." This clarification reduces ambiguity.

EXAMPLE 2: Correcting Misunderstandings

Initial Prompt: "Translate the following English text into French: 'He kicked the ball.'"

Analysis: If the model consistently mistranslates "kicked" as "licked," it indicates a misunderstanding. To correct this, the prompt can be iteratively adjusted by providing feedback and adding contextual cues: "Translate the following English text into French, ensuring that 'kicked' is correctly translated as 'a donné un coup de pied.'" The iterative process addresses this type of error.

EXAMPLE 3: Improving Completeness

Initial Prompt: "Describe the life of Leonardo da Vinci."

Analysis: The initial prompt lacks guidance on what aspects of da Vinci's life to include in the results, potentially leading to incomplete responses. Iteratively refine the prompt by specifying the scope, e.g., "Provide a comprehensive overview of Leonardo da Vinci's early life, artistic career, and inventions." This refinement ensures completeness.

EXAMPLE 4: Clarifying User Intent

Initial Prompt: "Find news articles about AI."

Analysis: The initial prompt may yield a broad range of AI-related news articles. If the user's intent is to find recent developments in AI ethics, the prompt can be iteratively improved with specificity: "Find news articles from the past month that discuss ethical concerns in AI technology." This adjustment aligns with the user's intent more effectively.

EXAMPLE 5: Contextual Consistency

Initial Prompt: "Continue the story from where it left off."

Analysis: This narrative prompt may lead to inconsistent or disjointed story-telling if the model lacks context from previous story segments. To maintain consistency, the prompt can be iteratively enhanced with contextual cues: "Continue the story from where it left off, where the protagonist was trapped in a mysterious cave." This refinement provides essential context.

In conclusion, strategies for iterating and improving prompts are integral to the process of prompt engineering. By clarifying specificity, incorporating feedback, adjusting scope, aligning with user intent, and maintaining contextual consistency, prompt engineers can continually enhance prompts to optimize model responses over time.

7

ADVANCED PROMPT ENGINEERING

Chapter 7 addresses advanced techniques that give users greater control over ChatGPT's responses. By mastering parameter tools like temperature, Top-p, max tokens, and incorporating conditional logic into prompts, we explore the frontiers of prompt engineering for dynamic and tailored interactions.

7.1 USING TEMPERATURE, TOP-P, AND MAX TOKENS FOR CONTROL

Fine-tuning the behavior of language models often requires nuanced control over response generation. In this section, we explore techniques such as temperature, top-p (nucleus sampling), and max tokens, which provide effective means to influence the creativity, coherence, and length of model responses. These techniques are instrumental for shaping the output of LLMs. The practical examples of prompts in this chapter were developed using ChatGPT application.

Temperature: Modulating Creativity

In the context of AI, particularly in language models like GPT, "temperature" refers to a parameter that controls the randomness of the model's responses. A higher temperature, such as 0.8, results in more creative and diverse responses, while a lower value, like 0.2, produces more deterministic and focused outputs. Temperature ranges from 0 to 2, with recommended default value of 1 for balance responses. Users can modify this parameter through OpenAI API.

EXAMPLE 1: Influencing Creativity

Prompt: "Write a poem about the ocean."

Analysis: By adjusting the temperature parameter, prompt engineers can fine-tune the creativity of the model's responses. A higher temperature might lead to imaginative and metaphor-rich poems, while a lower temperature can yield structured and conventional verses.

Top-p (Nucleus Sampling): Ensuring Relevance

Top-p is a technique that restricts the model's choices during token generation. It selects from the top-p probability distribution of words, ensuring that generated text remains relevant to the context. When the model is about to predict the next word or token in a sequence, instead of always choosing the most probable word, it considers a set of the most probable words. For example, if top-p is set to 0.9, the model will randomly pick the next word from a set of words that together have a cumulative probability of at least 90%. The value of Top-p is between 0 and 1. The higher value generates more text more random whereas a lower value generates word more deterministic. Typical values are set between 0.7 to 0.95.

EXAMPLE 2: Enhancing Relevance

Prompt: "Summarize the key findings of the latest climate change report."

Analysis: When using top-p sampling (e.g., top-p=0.7), the model is more likely to produce concise and contextually relevant summaries, as it focuses on the most probable words. This technique reduces the risk of irrelevant or verbose responses.

Max Tokens: Limiting Response Length

Max tokens is a parameter that restricts the length of model responses. It allows prompt engineers to enforce concise replies or set specific response lengths.

EXAMPLE 3: Controlling Response Length

Prompt: "Explain the process of photosynthesis in 100 words."

Analysis: By specifying max tokens (e.g., max tokens=100), prompt engineers can ensure that the model generates concise explanations within the defined word limit. This technique is particularly useful for tasks requiring brevity.

In conclusion, temperature, top-p (nucleus sampling), and max tokens are versatile tools for controlling the behavior of language models. By adjusting these parameters, prompt engineers can fine-tune the creativity of the AI's response, enhance its relevance, and limit the response length, effectively tailoring model outputs to meet specific requirements and optimize the user experience.

7.2   INCORPORATING CONDITIONAL LOGIC IN PROMPTS

The incorporation of conditional logic within prompts is a powerful technique for providing explicit instructions to guide model behavior. In this section, we explore how conditional statements and logic can be employed to influence and control the output of language models using research insights and practical examples.

Conditional Logic: Steering Model Responses

Conditional logic introduces a structured framework within prompts, enabling prompt engineers to specify conditions that should be met for desired responses. These conditions can range from simple constraints to complex decision trees.

EXAMPLE 1: Controlling Tone

Prompt: "Write a news article about recent climate change developments. If the tone is alarmist, provide counterarguments to promote balanced reporting."

Analysis: This prompt incorporates conditional logic by instructing the model to consider the tone of the response. If the tone seems alarmist, the model is guided to include counterarguments to ensure balanced reporting.

EXAMPLE 2: Customizing Output

Prompt: "Generate a product description for a smartphone. If the phone has a high-resolution camera, emphasize its photography capabilities; otherwise, focus on other features."

Analysis: Conditional logic in this prompt tailors the response to the smartphone's specific features. If a high-resolution camera is present, the model is directed to highlight photography capabilities; otherwise, it emphasizes other attributes.

EXAMPLE 3: Error Handling

Prompt: "Provide instructions for troubleshooting a Wi-Fi connectivity issue. If the user mentions interference from neighboring networks, suggest changing the channel; otherwise, advise checking the router settings."

Analysis: In this prompt, conditional logic handles different scenarios of Wi-Fi issues. If the user mentions interference, the model suggests changing the channel; otherwise, it provides advice on router settings.

EXAMPLE 4: Contextual Content Generation

Prompt: "Write a dialogue between two characters, Alice and Bob. If the conversation turns argumentative, steer it towards resolution and reconciliation."

Analysis: Conditional logic in narrative prompts ensures that even if the AI-generated dialogue initially turns argumentative, the model is guided to transition it towards resolution and reconciliation, maintaining a constructive narrative.

EXAMPLE 5: Multi-Turn Conversations

Prompt: "Engage in a conversation with the user. If they mention a specific topic, ask follow-up questions to gather more information; otherwise, introduce a general topic for discussion."

Analysis: In multi-turn conversations, conditional logic can manage user interactions. If the user mentions a specific topic, the model asks follow-up questions; otherwise, it initiates a general discussion.

In conclusion, incorporating conditional logic into prompts offers a structured means to steer and control model responses effectively. Whether for tone guidance, content customization, error handling, narrative control, or dynamic user interactions, conditional statements empower prompt engineers to provide clear instructions and fine-tune model behavior.

7.3 DYNAMIC PROMPTS FOR INTERACTIVE EXPERIENCES

Dynamic prompts introduce a level of interactivity and responsiveness to AI-generated content, enabling engaging and interactive experiences. In this section, we explore the concept of dynamic prompts and how they can be employed to create immersive and personalized interactions.

Dynamic Prompts: Fostering Engagement

Dynamic prompts are designed to respond to user input, adapt to changing contexts, and provide interactive and personalized experiences. They are particularly valuable in applications that require ongoing user interaction.

EXAMPLE 1: Interactive Fiction

Initial Prompt: "You are a detective investigating a murder mystery. Describe your first action."

Analysis: This example demonstrates the use of dynamic prompts in interactive fiction. The story progresses based on the user's decisions. For instance, if the user chooses to explore the crime scene, the prompt evolves accordingly: "You arrive at the crime scene to begin your investigation. What specific area or item catches your attention first?" Such adaptive prompting deepens user engagement by creating a responsive and evolving narrative experience.

EXAMPLE 2: Contextual Chatbots

AI Prompt: "Hello! How can I assist you today?"

Analysis: Dynamic prompts in chatbots enable personalized interactions. If a user mentions they want to book a hotel, the AI prompt adapts with the following response: "Great! I can help you with that. When and where would you like to book a hotel?" This approach tailors responses to user intent.

EXAMPLE 3: Game-Based Learning

AI Prompt: "Let's learn about history! Which historical period interests you?"

Analysis: Dynamic prompts in educational applications allow users to choose their learning path. If a user selects "Ancient Egypt," the AI adapts with the following response: "Excellent choice! Let's explore the fascinating world of

Ancient Egypt. What aspect would you like to learn about first?" This approach engages users in self-directed learning.

EXAMPLE 4: Personalized Content Recommendations

User Prompt: "Recommend a movie to watch."

Analysis: Dynamic prompts for content recommendations consider user preferences. If a user mentions a love for science fiction, the prompt adapts: "Sure! Based on your interest in science fiction, I recommend watching 'Blade Runner 2049.' What do you think?" This approach tailors suggestions to individual tastes.

EXAMPLE 5: Real-Time Collaboration

User Prompt: "Let's work on a collaborative story. Each of us will write one sentence to continue the narrative. I'll start: Once upon a time..."

Analysis: Dynamic prompts in collaborative storytelling adapt to user-contributed content. After a user's sentence, the prompt can respond: "Great beginning! Now, let's continue the story: 'in a magical forest where...'" This approach fosters real-time collaboration and creativity.

In conclusion, dynamic prompts introduce a new dimension of interactivity and responsiveness to AI-driven experiences. Whether for interactive fiction, chatbots, educational applications, content recommendations, or real-time collaboration, dynamic prompts empower prompt engineers to create engaging and personalized interactions that adapt to user input and preferences.

8

EFFECTIVE USE OF PROMPTS AND GPT-4 WITH PLUGINS

As the conversation between humans and AI models continues to evolve, plugins have become a potent tool in the design of prompts tailored for ChatGPT (more specifically, GPT-4). This chapter explores the multifaceted world of plugins, showing their potential to revolutionize industries and redefine user experiences. Whether you are an experienced AI engineer or a newcomer to the realm of prompt engineering, the insights offered here will help you to better utilize plugins and discover the full potential of AI-powered conversations.

8.1 AN INTRODUCTION TO PLUGINS

Plugins are a powerful addition to the GPT-4 ecosystem. They enhance the capabilities of the base model, allowing it to perform a wider range of tasks and provide more accurate answers. In this introductory section, we explore various plugins, unveiling their transformative potential in fine-tuning AI interactions. These ingenious tools enrich the user-AI conversational experience by introducing specialized functionalities. As we delve deeper into this information, you will gain an understanding of how plugins empower AI to excel in various domains, from language translation and sentiment analysis to content summarization. Through a close examination of real-world examples, we will uncover the versatility and ingenuity that plugins bring to the forefront of AI-driven conversations.

8.2 INTEGRATING PLUGINS FOR ENHANCED AI CONVERSATIONS

In this section, we delve into the intricacies of seamlessly integrating plugins to augment the capabilities of GPT-4 and make AI-driven conversations more sophisticated. We will explore the practical aspects of acquiring, implementing, and optimizing plugins to fine-tune AI interactions. As we navigate the process of integration, we will discuss how the synergy between prompts, GPT-4, and plugins can be used to craft dynamic and responsive conversations that address diverse needs and domains. Please refer to the screen-capture video provided with the accompanied source, for the updated implementation process and format for Plugins.

How To Get Plugins

Plugins are available with subscription of GPT-4. The following steps show how to install and use them.

1. **Subscription:** Ensure you have an active subscription to GPT-4. If you do not, visit the official OpenAI Web site (where GPT-4 is offered) and follow the subscription process.

2. **Accessing Plugin Section:** Once logged in, start a new chat session by clicking on the "+New chat" button. In the main area, hover your cursor over "GPT-4" and click on "Plugins."

3. **Browse & Choose:** Open the drop-down menu and click on "Plugin store." Here you will find a library of available plugins, ranging from specialized vocabularies to niche domain adaptations. Browse through them and read the descriptions to determine which ones will benefit you the most.

4. **Installation:** To install your desired plugin, click on "Install" icon. You can choose and install several plugins.

5. **Configuration:** To enable installed plugins, click on the drop-down menu and select the one you want to enable. Installed plugin icons will show up in a list at the top of the page. You can activate maximum three for each chat session. Follow the accompanying documentation or guidelines on the related Web site for each plugin tool to get more detail information and ensure they operate correctly.

6. **Usage:** Once installed and activated, the plugins will be employed by the GPT-4 or users can specify its usage by typing the name of desired plugin in the prompt.

7. **Updating & Management:** It is a good practice to check the "Plugins" section periodically for updates. Developers might release improvements or patches for their plugins, ensuring they remain compatible and optimized with the latest version of GPT-4.

8. **Uninstallation:** If you decide that you no longer need a particular plugin, navigate back to the "Plugin Store" section, and look for an "Uninstall" button. Clicking this button will remove the plugin from your GPT-4 instance.

Remember that while plugins can enhance the capabilities of GPT-4, they might also introduce specific biases or focus areas. Always test new plugins to ensure they meet your expectations and needs.

EXAMPLE 1: Wolfram Plugin

Plugins generally augment or adjust the responses from GPT-4. For example, after integrating the Wolfram plugin, GPT-4 becomes proficient in handling complex mathematical or scientific inquiries. If you prompt GPT-4 with an enabled Wolfram plugin with "Calculate the integral *arcsin(x)* as indefinite integral," the Wolfram-enhanced GPT-4 would seamlessly retrieve and present a precise answer, as shown in Figure 8.1.

Used Wolfram ⌄

The indefinite integral of $\arcsin(x)$ with respect to x is given by:

$$\int \arcsin(x)\, dx = \sqrt{1 - x^2} + x \arcsin(x) + C$$

where C is the constant of integration.

Here are the plots of the integral:

1. For x in the range $[-1, 1]$:

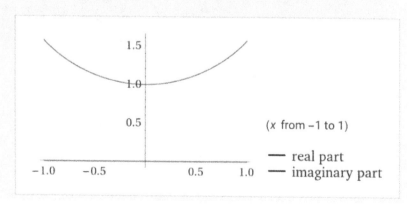

FIGURE 8.1 A Wolfram-enhanced GPT-4 response to the prompt "Calculate the integral *arcsin(x)* as indefinite integral."

To see the code, click on the drop-down menu from the "Used Wolfram" icon. The Figure 8.2 shows the code used for this example.

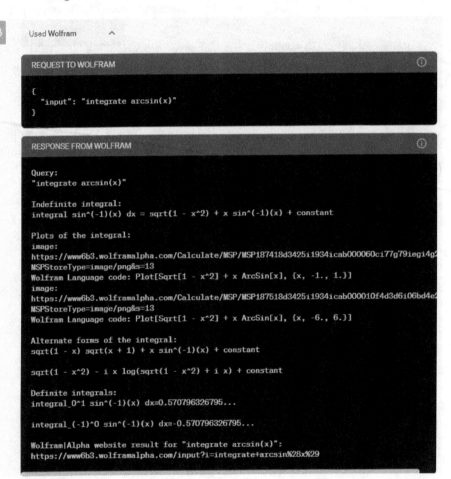

Used Wolfram ⌃

REQUEST TO WOLFRAM ⓘ

```
{
  "input": "integrate arcsin(x)"
}
```

RESPONSE FROM WOLFRAM ⓘ

```
Query:
"integrate arcsin(x)"

Indefinite integral:
integral sin^(-1)(x) dx = sqrt(1 - x^2) + x sin^(-1)(x) + constant

Plots of the integral:
image:
https://www6b3.wolframalpha.com/Calculate/MSP/MSP187418d3425i1934icab000060ci77g79iegi4g
MSPStoreType=image/png&s=13
Wolfram Language code: Plot[Sqrt[1 - x^2] + x ArcSin[x], {x, -1., 1.}]
image:
https://www6b3.wolframalpha.com/Calculate/MSP/MSP187518d3425i1934icab000010f4d3d6i06bd4e
MSPStoreType=image/png&s=13
Wolfram Language code: Plot[Sqrt[1 - x^2] + x ArcSin[x], {x, -6., 6.}]

Alternate forms of the integral:
sqrt(1 - x) sqrt(x + 1) + x sin^(-1)(x) + constant

sqrt(1 - x^2) - i x log(sqrt(1 - x^2) + i x) + constant

Definite integrals:
integral_0^1 sin^(-1)(x) dx≈0.570796326795...

integral_(-1)^0 sin^(-1)(x) dx≈-0.570796326795...

Wolfram|Alpha website result for "integrate arcsin(x)":
https://www6b3.wolframalpha.com/input?i=integrate+arcsin%28x%29
```

FIGURE 8.2 The code used for the Wolfram example.

EXAMPLE 2: Prompt Perfect Plugin

Plugins generally augment or adjust the responses from GPT-4. For example, after integrating the Prompt Perfect plugin, GPT-4 becomes proficient in handling prompts by rephrasing them to more complex and effective ones. In other words, it automates prompt engineering application. If you prompt GPT-4 with the prompt "Prompt Perfect. What is climate change?," the Prompt-Perfect-enhanced GPT-4 would seamlessly retrieve and present a very precise answer. See Figure 8.3.

```
Used Prompt Perfect        ^

REQUEST TO PROMPT PERFECT                                    (i)

{
  "text": "what is climate change?"
}

RESPONSE FROM PROMPT PERFECT                                 (i)

{
  "rephrased": {
    "text": "Can you please provide a comprehensive explanation of climate change,
including its causes, impacts, and potential solutions?"
  }
}
```

FIGURE 8.3 The original prompt (top) and the engineered prompt (bottom).

EXAMPLE 3: KeyMate Plugin

The KeyMate plugin seamlessly integrates the capabilities of ChatGPT/GPT-4 with Google Search, introducing a persistent memory feature to chat interactions. This innovative approach ensures that conversations are not only informative but also retain context over time, enhancing the user experience and providing a more cohesive interaction. Furthermore, while OpenAI has stated that ChatGPT is informed by data up until 2021, the inclusion of the KeyMate plugin offers users a pathway to bypass this constraint to a certain degree. Figure 8.4 shows an interaction with ChatGPT enhanced with the KeyMate plugin.

Used **KeyMate.AI Search**   ⌄

Used **KeyMate.AI Search**   ⌄

Used **KeyMate.AI Search**   ⌄

Tomorrow's weather forecast for Vancouver on 14th September 2023 is as follows:

* **Condition**: Mainly sunny.
* **High Temperature**: 26°C.
* **Low Temperature at Night**: 12°C.
* **UV Index**: 6 or high.
* **Wind**: Expected to become southwest at 20 km/h in the afternoon.

FIGURE 8.4 KeyMate-enhanced ChatGPT's response to the prompt "Tomorrow's weather forecast in Vancouver, SI units. Express the exact date. KeyMate."

An increasing number of plugins are being developed for GPT-4. Notable examples include Expedia, Instacart, Kayak, Chatwithpdf, Visla, and Smart Slides. Users can conveniently access the catalog of available plugins via the Plugin Store within GPT-4 each time they initiate a new chat.

8.3 BENEFITS OF USING PLUGINS

The integration of plugins with ChatGPT offers a multitude of advantages, enhancing its capabilities and making it more adaptable to various tasks. Some of the primary benefits include the following:

- **Enhanced Knowledge Base**: By utilizing external data sources, Chat-GPT ensures that the answers it provides are grounded in the most up-to-date information available. This ensures accuracy and relevance in its responses.
- **Dynamic Computations**: With plugins, ChatGPT is not just a static information provider. It can perform real-time calculations, offering solutions to mathematical, scientific, and other computational queries instantly.
- **Integration with Third-party Services**: This is perhaps one of the most exciting benefits. By integrating with various third-party services, ChatGPT can perform a wide range of tasks. Whether it is scheduling an appointment, obtaining the latest news, or controlling smart devices in a connected home, the possibilities are vast and continually expanding.
- **Customization and Flexibility**: Plugins allow users to tailor ChatGPT's capabilities to their specific needs. Whether it is for business analytics, educational purposes, or personal tasks, plugins can be added or removed to customize the user experience.
- **Continuous Learning and Improvement**: With the integration of plugins, ChatGPT can continuously learn and update its knowledge base. This ensures that the AI remains at the forefront of technological advancements and can provide users with the most current and relevant information.
- **Growing Library of Plugins**: As the community of developers and users expands, so does the library of available plugins. This ensures that Chat-GPT remains adaptable to emerging trends, technologies, and user needs. With a growing library, users have access to a broader range of functionalities and can continuously discover new ways to utilize ChatGPT.

Using plugins transforms ChatGPT from a mere conversational AI to a versatile tool capable of assisting users in numerous ways.

8.4 POPULAR CHATGPT/GPT-4 PLUGINS

Plugins are an integral part of enhancing the capabilities of various platforms. ChatGPT, in particular, has seen a surge in the development and adoption of plugins that augment its functionalities. These plugins range from Web browsing tools to dynamic computation modules, each designed to cater to specific user needs.

Currently, there are more than 200 plugins available on the ChatGPT Plus/ GPT-4 that aid users in their daily tasks across varying fields. Whether it is for educational purposes, research, cooking, or entertainment, there is likely a plugin tailored for the desired purpose. Here are some of the plugins that have gained popularity among ChatGPT Plus users:

1. **Ask Your PDF**: Tailored for researchers, this plugin allows users to directly ask questions and get answers from the information written in PDFs.

2. **Link Reader**: Can read the content and extract information from all kinds of Web links, including Web pages, PDFs, images, and more.

3. **There's An AI For That**: Offers a database of AI tools for various purposes, including image editing and PDF conversion.

4. **Prompt Perfect**: Assists users in crafting useful prompts for AI chatbots.

5. **World News**: Provides a well-organized list of news articles in multiple languages and source links.

6. **Stories**: Allows users to engage in the art of storytelling, generating captivating stories based on provided prompts.

7. **Show Me Diagrams**: Enables users to swiftly generate visual diagrams accompanied by text explanations.

8. **MixerBox OnePlayer**: A comprehensive music compilation tool that curates playlists based on user preferences. MixerBox Calendar and MixerBox FindPlug are more plugins from this developer.

9. **VoxScript**: Obtains a video transcript and lets users quickly extract useful information.

10. **Chat with PDF**: Similar to "Ask Your PDF," it comprehends textbooks, handouts, and presentations.

11. **ScholarAI**: Grants users access to a database of scholarly articles and academic research.

12. **VideoInsights**: Analyzes video content to generate summaries, action items, and key highlights.

13. **KeyMate.AI Search**: Enhances your knowledge base by searching the Internet for the latest information on diverse subjects.

14. **WebPilot**: Enables users to interact with, extract specific information from, or modify the content of Web sites.

15. **Wolfram**: It allows ChatGPT to perform complex calculations and access curated data.

16. **Content Rewriter**: With this plugin, ChatGPT can rewrite content from a given URL in its own words.

These plugins, among many others, showcase the versatility and adaptability of ChatGPT, making it a powerful tool in various domains. Users may want to check the growing list of plugins at the GPT-4 Plugin Store.

CHAPTER

9

REAL-WORLD APPLICATIONS

Chapter 9 examines the myriad ways AI and prompt engineering technologies can be used in real-world applications. We discuss how prompt engineering enhances both commercial and private projects, from customer support chatbots streamlining user inquiries to creative writing assistance fueling novel narratives. We demonstrate the powerful impact of AI in domains like customer support and business, content generation, teaching and learning, and engineering applications, and explore how prompt engineering can improve the relationship between human intent and machine response. You will be able to observe AI's transformative power in the everyday world, where theory becomes reality and innovation thrives.

9.1 APPLYING PROMPT ENGINEERING IN CUSTOMER SUPPORT

Prompt engineering plays a pivotal role in customer support, elevating the quality and efficiency of interactions between support agents and customers. By crafting well-constructed prompts, customer support teams can ensure that AI models like GPT-4 provide accurate and helpful responses, leading to quicker issue resolution and improved customer satisfaction.

Enhancing Customer Support with Prompt Engineering

Prompt engineering in customer support enables organizations to streamline their support operations, delivering prompt responses, accurate information, and improved user experiences.

EXAMPLE 1: Streamlining Technical Support

Prompt: "As a customer support agent, troubleshoot a user's connectivity issues with their Wi-Fi router. User reports frequent disconnections."

Analysis: In this example, the prompt clearly defines the role, goal, and context for the AI model. It instructs the AI to act as a customer support agent and focus on resolving the user's connectivity issues with their Wi-Fi router, with the understanding that the user has been experiencing frequent disconnections. Such prompts enable customer support agents to efficiently address specific technical issues, leading to smoother and more effective support interactions.

EXAMPLE 2: Billing and Account Assistance

Prompt: *"As a customer support agent, assist a user with billing and account inquiries. User reports discrepancies in their recent billing statement."*

Analysis: In this instance, the prompt defines the role of the AI model as a customer support agent, specifies the goal as assisting with billing and account inquiries, and provides context by highlighting the user's concern regarding discrepancies in their recent billing statement. This well-crafted prompt empowers the AI to focus on addressing the user's specific issues, facilitating a more productive interaction between the customer support agent and the user.

EXAMPLE 3: Product Recommendations

Prompt: *"As a customer support agent, offer tailored product recommendations. User seeks suggestions for purchasing a new laptop within a specific budget range and for particular use cases."*

Analysis: In this example, the prompt assigns the role of a customer support agent to the AI and identifies the goal of offering tailored product recommendations. The context specifies that the user is seeking advice on purchasing a new laptop within a particular budget range and for specific use cases. This well-crafted prompt empowers the AI to provide relevant and customized product suggestions, enhancing the user's shopping experience and adding value to the customer support interaction.

In conclusion, prompt engineering in customer support helps agents to deliver efficient, accurate, and user-centric interactions. By employing RGC inquiry prompts, identity verification prompts, preference elicitation prompts, order information prompts, and policy clarification prompts, organizations can elevate their customer support capabilities, providing prompt solutions and enhancing user satisfaction.

9.2 GENERATING CONTENT FOR SOCIAL MEDIA AND MARKETING

Social media and marketing can benefit from the potential of AI-driven content generation. In this section, we delve into how prompt engineering and language models are revolutionizing content creation for social media platforms and marketing campaigns. Drawing from research and practical examples, we unveil the pivotal role of AI in crafting engaging and impactful content.

Transforming Social Media with AI-Powered Content

Social media platforms thrive on fresh, engaging content. AI, powered by prompt engineering, has emerged as a game-changer in this arena, offering scalable content generation with a human touch.

EXAMPLE 1: Social Media Posts

Prompt: "As a content creator, craft compelling social media posts. Create engaging posts to promote a new product launch for a tech company."

Analysis: In this example, the prompt assigns the role of a content creator to the AI and sets the goal as crafting compelling social media posts. The context specifies that the posts should focus on promoting a new product launch for a tech company. This well-constructed prompt guides the AI to generate attention-grabbing and relevant social media content, contributing to a successful marketing campaign.

EXAMPLE 2: Hashtag Campaigns

Prompt: "As a social media manager, develop an impactful hashtag campaign. Create the hashtag and associated content to generate buzz around an upcoming product release."

Analysis: In this example, the prompt assigns the role of a social media manager and identifies the goal of developing an impactful hashtag campaign. The context specifies that the campaign should aim to create a trending hashtag and associated content to generate excitement and anticipation for an upcoming product release. This prompt provides clear direction to the AI, enabling it to generate hashtag ideas and campaign content that resonate with the target audience and maximize engagement.

EXAMPLE 3: Product Descriptions

Prompt: "As a copywriter, create persuasive product descriptions. Generate captivating and informative product descriptions for a new line of tech gadgets."

Analysis: In this example, the prompt assigns the role of a copywriter to the AI and identifies the goal of creating persuasive product descriptions. The context specifies that the task involves generating captivating and informative product descriptions for a new line of tech gadgets. This carefully crafted prompt guides the AI to produce product descriptions that interest potential buyers and provide valuable details about the featured tech gadgets, ultimately enhancing the e-commerce shopping experience.

In summary, AI-powered content generation, driven by prompt engineering, is transforming social media engagement and marketing campaigns. Through engagement-focused prompts, trend-integrated prompts, feature-centric prompts, experience-driven prompts, and persuasive prompts, organizations can harness the creative potential of AI to craft compelling content that resonates with audiences and drives marketing success.

9.3 EDUCATIONAL USE CASES: TEACHING AND LEARNING WITH CHATGPT

Education is ready for a digital revolution, and AI, particularly chatbots driven by prompt engineering, is at the forefront of this transformation. In this section, we explore how AI can revolutionize teaching and learning, drawing from research findings and real-world examples to illuminate its pivotal role in education.

Transforming Education with AI-Powered Chatbots

AI-driven chatbots are reshaping education by providing personalized, accessible, and interactive learning experiences for students of all ages.

EXAMPLE 1: Homework Assistance

Prompt: "As a virtual tutor, aid a student in solving a calculus problem. The student requires step-by-step guidance on finding the derivative of a complex trigonometric function."

Analysis: In this example, the prompt assigns the role of a virtual tutor to the AI and defines the goal as aiding a student in solving a calculus problem. The context specifies that the student needs step-by-step guidance on finding the derivative of a complex trigonometric function. This thoughtfully crafted prompt guides the AI to provide educational support tailored to the specific academic needs of the student, facilitating a more effective and engaging learning experience. Prompts can be followed by further queries to help student to articulate their understanding of a problem.

EXAMPLE 2: Language Learning

Prompt: "As a programming instructor, teach the basics of Python programming. Students are beginners and seek an introduction to Python, including variables, data types, and simple programs."

Analysis: In this example, the prompt assigns the role of programming instructor to the AI and defines the goal as teaching the basics of Python programming. The context specifies that the students are beginners who seek an introduction to Python, including concepts such as variables, data types, and simple programs. This carefully crafted prompt guides the AI to provide an effective and structured introduction to Python programming, catering to the students' specific learning needs and objectives.

EXAMPLE 3: Concept Clarification

Prompt: "As a subject matter expert, explain the principles of quantum mechanics. Students seek a clear and concise explanation of quantum mechanics concepts, including wave-particle duality and quantum states."

Analysis: In this example, the prompt assigns the role of a subject matter expert to the AI and defines the goal as explaining the principles of quantum mechanics. The context specifies that the students are seeking a clear and concise explanation of quantum mechanics concepts, with a focus on understanding topics like wave-particle duality and quantum states. This thoughtfully crafted prompt guides the AI to provide an informative and comprehensible explanation, facilitating the student's grasp of complex scientific concepts.

EXAMPLE 4: Test Preparation

Prompt: "As an exam coach, assist a student in practicing algebraic equations. The student needs guided practice on solving algebraic equations involving variables, fractions, and simplification for an upcoming math exam."

Analysis: In this example, the prompt assigns the role of an exam coach to the AI and defines the goal as assisting a student in practicing algebraic equations. The context specifies that the student requires guided practice in solving algebraic equations that involve variables, fractions, and simplification, all in preparation for an upcoming math exam. This carefully crafted prompt guides the AI to provide tailored exercises and explanations to help the student improve their algebraic problem-solving skills and excel in their examination.

EXAMPLE 5: Capstone Project Report

Prompt: "As an engineering advisor, provide guidance on the capstone project report. Students are working on a capstone project report detailing the design, construction, and testing of a small wind turbine for renewable energy generation."

Analysis: In this example, the prompt assigns the role of an engineering advisor to the AI and identifies the goal as providing guidance on the capstone project report. The context specifies that the student team is working on a capstone project report that details the design, construction, and testing of a small wind turbine for renewable energy generation. This thoughtfully crafted prompt guides the AI to offer expert advice, insights, and feedback to help the student successfully complete their capstone project and report on the innovative design of a small wind turbine.

In conclusion, AI-driven chatbots, guided by prompt engineering, are reshaping education by providing personalized and interactive learning experiences. Whether it is homework assistance, language learning, concept clarification, test preparation, or tutoring, these chatbots offer valuable support and foster skill development for students of all levels. As the education industry continues to evolve, AI-powered chatbots can become indispensable tools in the pursuit of knowledge.

9.4 BUSINESS APPLICATIONS

AI-powered chatbots and prompt engineering in business are causing profound transformations. This section explores the myriad ways in which these technologies are revolutionizing business operations, decision-making, and customer interactions using research insights and practical examples.

Enhancing Business Operations with AI Chatbots

Businesses are increasingly turning to AI chatbots to streamline operations, reduce costs, and improve customer service.

EXAMPLE 1: Customer Service Chatbots

Prompt: "As a chatbot developer, give guidance to create an effective customer service chatbot. The business seeks to implement a chatbot to handle customer inquiries and support requests in real-time."

Analysis: In this example, the prompt assigns the role of chatbot developer to the AI and identifies the goal as creating an effective customer service chatbot. The context specifies that the business is interested in implementing a chatbot to handle customer inquiries and support requests in real-time. This carefully crafted prompt guides the AI to focus on developing a chatbot solution tailored to the specific needs and objectives of the business, streamlining customer support processes, and enhancing the overall customer experience.

EXAMPLE 2: Sales Assistance

Prompt: "As a sales support AI, assist customers in making purchase decisions. Customers are seeking guidance and recommendations for selecting the right products based on their preferences and needs."

Analysis: In this example, the prompt assigns the role of a sales support AI to the AI and identifies the goal as assisting customers in making purchase decisions. The context specifies that customers are seeking guidance and recommendations for selecting the right products based on their preferences and needs. This thoughtfully crafted prompt guides the AI to offer personalized product suggestions and assistance, helping customers make informed purchasing decisions and contributing to increased sales and customer satisfaction.

EXAMPLE 3: Lead Generation

Prompt: "As a lead generation AI, identify and qualify potential leads. The business aims to leverage AI to identify and qualify potential customers who have expressed interest in their products or services. Use the Outbound Marketing method."

Analysis: In this example, the prompt assigns the role of a lead generation AI to the AI and identifies the goal as identifying and qualifying potential leads. The context specifies that the business is looking to harness AI to identify and assess potential customers who have shown interest in their products or services. This carefully crafted prompt guides the AI to focus on processes that help identify and categorize potential leads efficiently, ultimately contributing to the growth and success of the business's sales efforts.

EXAMPLE 4: Employee Onboarding

Prompt: "As an onboarding assistant AI, streamline the employee onboarding process. Business aims to leverage AI to automate and enhance the employee onboarding experience for new hires."

Analysis: In this example, the prompt assigns the role of an onboarding assistant to the AI and identifies the goal as streamlining the employee onboarding process. The context specifies that the business is looking to use AI to automate and improve the overall employee onboarding experience for new hires. This thoughtfully crafted prompt guides the AI to focus on developing solutions that expedite and simplify the onboarding process, allowing new employees to seamlessly integrate into the organization, ultimately leading to increased efficiency and employee satisfaction.

EXAMPLE 5: Data Analysis

Prompt: "As a data analysis AI, analyze and extract insights from complex datasets. Business seeks to leverage AI to process and interpret large volumes of data, extracting valuable insights for informed decision-making."

Analysis: In this example, the prompt assigns the role of a data analysis AI to the AI and identifies the goal as analyzing and extracting insights from complex datasets. The context specifies that the business is interested in using AI to handle the processing and interpretation of large volumes of data, with the objective of extracting valuable insights to support informed decision-making. This carefully crafted prompt guides the AI to focus on data analytics tasks, enabling the business to make data-driven decisions and gain a competitive edge.

In conclusion, AI chatbots empowered by prompt engineering are motivating efficiency and innovation across diverse business sectors. Whether it is customer service, sales, lead generation, employee onboarding, or data analysis, these chatbots are enhancing operations, improving decision-making, and elevating customer experiences. As businesses continue to embrace AI-driven solutions, chatbots are poised to become indispensable allies in achieving operational excellence.

9.5 TECHNICAL AND ENGINEERING APPLICATIONS

The role of prompt engineering in the technical and engineering domains has emerged as a transformative force. This section examines the multifaceted applications of prompt engineering within the realms of technology and engineering. From aiding in product design and structural analysis to streamlining manufacturing processes and fostering innovation in materials science, well-constructed prompts are indispensable tools for professionals and researchers alike. Here, we explore how the art of prompt engineering has been harnessed to drive progress, optimize operations, and spur innovation across various technical and engineering disciplines. Through a series of illustrative examples, we uncover the impact of this field on problem-solving, idea generation, and the pursuit of excellence in technical and engineering endeavors. This section examines the intersection of prompt engineering and technological innovation, where precision in language meets the intricacies of complex systems, fostering creativity, efficiency, and advancement.

EXAMPLE 1: Product Design

Prompt: "As a product designer, conceptualize a sustainable urban transportation solution. Design an eco-friendly and efficient mode of urban transportation that addresses the challenges of congestion and environmental impact."

In this example, the prompt defines the role of the AI as a product designer and identifies the goal as conceptualizing a sustainable urban transportation solution. The context specifies that the design should focus on creating an eco-friendly and efficient mode of urban transportation while addressing the pressing challenges of congestion and environmental impact. This carefully crafted prompt guides the AI to generate design ideas that align with sustainability goals, contributing to the evolution of urban transportation solutions.

As another example, consider the following prompt:

Prompt: "As an automotive component designer, innovate a high-performance car brake system component. Design a novel and efficient brake caliper component that enhances braking performance, durability, and safety in high-performance sports cars."

In this prompt, the AI is assigned the role of an automotive component designer with the goal of innovating a high-performance car brake system component. The context specifies that the design challenge involves creating a novel brake caliper component. This component should not only enhance braking performance but also improve durability and safety, particularly for use in high-performance sports cars. This well-constructed prompt guides the AI to focus on designing a cutting-edge brake system component, pushing the boundaries of automotive engineering for enhanced performance and safety.

EXAMPLE 2: Structural Analysis

Prompt: "As a structural engineer, evaluate the load-bearing capacity of a steel bridge main girder. Perform a comprehensive structural analysis of the girder, focusing on load bearing, stress distribution, and potential areas of reinforcement."

In this example, the prompt assigns the role of structural analyst to the AI and identifies the goal as evaluating the load-bearing capacity of a steel bridge girder. The context specifies that the analysis should be comprehensive, with a particular focus on load-bearing capacity, stress distribution, and identifying areas that may require reinforcement. This well-constructed prompt guides the AI to conduct a rigorous structural analysis, ensuring the safety and longevity of the bridge while adhering to engineering standards and best practices.

EXAMPLE 3: Computer-Aided Design (CAD)

Prompt: "As a CAD designer, create the list of main features for a concept design for an ergonomic office chair. Develop a concept design for an ergonomic office chair that prioritizes user comfort, support, and aesthetic appeal while adhering to ergonomic principles."

Analysis: In this example, the prompt assigns the role of a CAD designer to the AI and identifies the goal as creating a concept design for an ergonomic office chair. The context specifies that the design should prioritize user comfort, support, and aesthetic appeal while adhering to ergonomic principles. This carefully constructed prompt guides the AI to generate a CAD design that not only encourages an innovative approach, but also ensures that the final product meets the highest standards of ergonomic design, enhancing user experience and productivity in office environments.

EXAMPLE 4: Material Engineering and Science

Prompt: "As a material scientist/engineer, identify and suggest a new material with exceptional thermal conductivity. Conduct research to discover a novel material with superior thermal conductivity properties for applications in electronics cooling and heat management."

Analysis: In this example, the prompt assigns the role of a material scientist/ engineer to the AI and identifies the goal as discovering a new material with exceptional thermal conductivity. The context specifies that the research should focus on finding a novel material with superior thermal conductivity properties, particularly for use in electronics cooling and heat management applications. This thoughtfully crafted prompt guides the AI to explore materials science, pushing the boundaries of thermal conductivity research to uncover materials that could revolutionize heat management solutions in various industries.

EXAMPLE 5: Environmental and Energy Engineering

Prompt: "As an engineer, develop an energy-efficient HVAC system for a green building. Design an HVAC system that maximizes energy efficiency and minimizes environmental impact in a LEED-certified green building project."

Analysis: In this example, the prompt assigns the role of engineer to the AI and identifies the goal as developing an energy-efficient HVAC (Heating, Ventilation, and Air Conditioning) system. The context specifies that the design should prioritize energy efficiency and environmental sustainability, particularly in the context of a LEED-certified green building project. This carefully crafted prompt guides the AI to design an HVAC system that aligns with green building standards, reducing energy consumption and contributing to a more sustainable and eco-friendlier environment.

EXAMPLE 6: Optimization and Automation

Prompt: "As a manufacturing automation specialist, implement an automated quality control system for a food packaging plant. Develop an automated solution that ensures precise quality checks, minimizes product waste, and enhances overall operational efficiency in a food packaging facility."

Analysis: In this prompt, the AI is assigned the role of a Manufacturing Automation Specialist, and the goal is to implement an automated quality control system. The context specifies that the focus should be on creating an automated solution that ensures precise quality checks, minimizes product waste, and improves the overall operational efficiency of a food packaging facility. This prompt guides the AI to explore automation technologies and solutions to optimize the food packaging process, ensuring high-quality products and reduced operational costs.

10

FUTURE TRENDS AND ETHICAL CONSIDERATIONS

The future of prompt engineering is intertwined with the evolution of AI models and techniques. We discuss emerging models, such as hybrid architectures and multi-modal approaches, that can enhance AI-generated content. By understanding these advancements, we prepare to leverage cutting-edge technologies in prompt engineering.

10.1   THE EVOLVING LANDSCAPE OF AI-GENERATED CONTENT

The tools involved in AI-generated content are in a state of transformation, marked by rapid advancements. This section discusses the tools needed for AI-generated content, exploring its impact on various domains and reflecting on its future. Drawing from research insights and real-world examples, we examine the dynamic nature of AI-generated content.

However, it is essential to recognize the inherent complexity in forecasting the future of AI advances. The pace of innovation in AI often defies prediction, as breakthroughs can emerge unexpectedly, disrupting established paradigms. The multifaceted nature of AI, spanning natural language processing, computer vision, and other domains, further compounds this challenge. Moreover, the ethical and societal implications of AI progress add layers of uncertainty to the equation.

Despite these complexities, our aim is to provide a forward-looking perspective on the evolving landscape of AI-generated content. By examining current trends and anticipating potential trajectories, we hope to empower readers with the insights needed to navigate this dynamic field and harness its transformative potential for the benefit of various industries and society at large.

The Unfolding Revolution of AI-Generated Content

AI-generated content is becoming increasingly pervasive, infiltrating diverse sectors and reshaping the way we create, interact with, and consume information.

EXAMPLE 1: Creative Writing

Prompt: "As a poet, compose a haiku that captures the beauty of a cherry blossom in spring, Craft a three-line poem that evokes the ephemeral nature of nature's artistry and the renewal of life with the changing seasons."

Analysis: In this example, the prompt assigns the AI the role of a poet and tasks it with composing a haiku, a traditional form of Japanese poetry. The goal is to create a three-line poem that beautifully encapsulates the essence of a cherry blossom in spring. The context emphasizes the haiku's purpose: to convey the fleeting yet profound beauty of nature's artistry and the cyclical renewal of life as the seasons change. This prompt guides the AI to generate a poetic masterpiece that paints a vivid picture of nature's wonders.

EXAMPLE 2: Content Generation

Prompt: "As a content creator, compose an informative article on sustainable agriculture practices. Write a well-researched article that highlights the importance of sustainable farming methods, their environmental benefits, and their potential to transform the future of agriculture."

Analysis: In this example, the prompt assigns the role of content creator to the AI and identifies the goal as composing an informative article. The context specifies that the article should focus on sustainable agriculture practices, emphasizing their significance, environmental advantages, and potential to revolutionize the agricultural industry. This well-constructed prompt guides the AI to generate a comprehensive and engaging article that educates readers on the topic of sustainable farming.

EXAMPLE 3: Personalization

Prompt: "As a user experience designer, design a personalized news recommendation system. Develop an algorithm that assembles news articles by considering each user's interests, browsing history, and preferences, enhancing their news consumption experience."

Analysis: In this example, the prompt assigns the role of a user experience designer to the AI and identifies the goal as designing a personalized news recommendation system. The context specifies that the focus should be on creating an algorithm that tailors news article recommendations to individual users, taking into account their interests, browsing history, and preferences. This well-constructed prompt guides the AI to develop a system that enhances the user's news consumption experience by delivering content that aligns with their unique tastes and interests.

EXAMPLE 4: Marketing and Advertising

Prompt: "As a marketing strategist, generate a series of engaging social media advertisements for a new line of eco-friendly products. Create visually appealing and persuasive ad content that highlights the products' sustainability, quality, and their positive impact on the environment."

Analysis: In this example, the prompt assigns the role of a marketing strategist to the AI and identifies the goal as generating a series of engaging social media advertisements. The context specifies that the focus should be on creating visually appealing and persuasive ad content that showcases a new line of eco-friendly products. The ads should emphasize the products' sustainability, quality, and the positive impact they have on the environment. This well-structured prompt guides the AI to craft advertisements that effectively communicate the product's unique selling points to potential customers.

EXAMPLE 5: Language Translation

Prompt: "As a language translator, translate a complex legal document from English to Spanish. Ensure accuracy, consistency, and adherence to legal terminology, providing a seamless and reliable translation for legal professionals."

Analysis: In this example, the prompt assigns the role of a language translator to the AI and identifies the goal as translating a complex legal document from English to Spanish. The context specifies that the focus should be on ensuring accuracy, consistency, and adherence to legal terminology, delivering a seamless and reliable translation service for legal professionals. This well-constructed prompt guides the AI to perform a high-quality translation, maintaining the document's legal integrity and making it accessible to a Spanish-speaking audience.

As AI continues to change how content is generated, the future holds promise and challenges alike. The ethical considerations surrounding AI-generated content, such as plagiarism and misinformation, necessitate vigilant monitoring. However, the potential for creative collaboration, personalized experiences, and streamlined content creation positions AI-generated content as a transformative force in our information-driven world.

10.2   ADDRESSING BIAS AND FAIRNESS IN PROMPTS AND RESPONSES

As AI systems become increasingly intertwined with human interactions, addressing bias and ensuring fairness in prompts and responses is important. This section delves into the crucial issue of bias in AI-generated content, exploring the challenges it poses and the strategies to mitigate it. Drawing from research findings and real-world examples, we discuss bias mitigation in prompt engineering.

The Perils of Bias in AI-Generated Content

Bias in AI-generated content can perpetuate stereotypes, reinforce discrimination, and erode trust in AI systems. It is essential to recognize and rectify bias to ensure equitable and inclusive human-AI interactions.

EXAMPLE 1: Political Bias in Information Retrieval

Prompts: "As an AI ethics analyst, develop a prompt template for generating neutral news summaries. Craft a standardized prompt format that ensures balanced and unbiased news summaries, free from political slant or editorial influence, thus promoting fair and accurate information retrieval."

Analysis: In this example, the prompt assigns the AI the role of an AI ethics analyst and identifies the goal as developing a prompt template. The context specifies that the focus should be on creating a standardized prompt format that guarantees balanced and unbiased news summaries. The aim is to ensure that the generated summaries are free from political slant or editorial influence, promoting fair and accurate information retrieval for users seeking news content. This prompt guides the AI to contribute to mitigating political bias in information retrieval systems.

EXAMPLE 2: Racial and Ethnic Bias in Translation

Prompt: As an AI translator, create a prompt framework for culturally sensitive translation. Develop a set of guidelines and prompt templates that prioritize culturally nuanced and respectful translations, ensuring that the AI translation system avoids perpetuating racial or ethnic bias."

Analysis: In this example, the prompt assigns the role of an AI translator to the AI and identifies the goal as creating a prompt framework. The context specifies that the focus should be on developing guidelines and prompt templates that prioritize culturally sensitive translations. The aim is to ensure that the AI translation system avoids perpetuating racial or ethnic bias in its translations. This well-constructed prompt guides the AI to contribute to fair and respectful cross-cultural communication by addressing bias in translation outputs.

EXAMPLE 3: Bias Amplification in Content Generation

Prompt: "As a content curator, design prompts that prevent bias amplification. Create a series of prompt templates that promote balanced, inclusive, and unbiased content generation, ensuring that the AI system does not inadvertently magnify or perpetuate existing biases in its responses."

Analysis: In this example, the prompt assigns the AI the role of a content curator and identifies the goal as designing prompts to prevent bias amplification. The context specifies that the focus should be on creating a series of prompt templates that prioritize balanced, inclusive, and unbiased content generation. The aim is to ensure that the AI system does not inadvertently magnify or perpetuate existing biases in its responses. This prompt guides the AI to contribute to content generation that promotes fairness and inclusivity while minimizing the risk of bias amplification.

EXAMPLE 4: Bias Detection and Redirection

Prompt: "As an AI bias analyst, create prompts for bias detection and redirection. Develop a set of prompt templates that empower the AI system to identify biased content in responses and provide corrections or alternative perspectives to ensure a fair and unbiased exchange of information."

Analysis: In this example, the prompt assigns the role of an AI bias analyst to the AI and identifies the goal as creating prompts for bias detection and redirection. The context specifies that the focus should be on developing prompt templates that enable the AI system to identify biased content in its responses and take corrective actions, such as providing corrections or alternative perspectives. The aim is to ensure a fair and unbiased exchange of information, mitigating the impact of bias in AI-generated content. This prompt guides the AI to contribute to the detection and redirection of biased content for more balanced and inclusive communication.

EXAMPLE 5: Bias-Responsive Systems

Prompt: "As an AI system developer, construct prompts for building bias-responsive systems. Develop a set of prompt templates that allow the AI system to adapt and respond to feedback on bias, fostering continuous improvement and ensuring that the system actively addresses and mitigates bias in its outputs."

Analysis: In this example, the prompt assigns the role of an AI system developer to the AI and identifies the goal as constructing prompts for building bias-responsive systems. The context specifies that the focus should be on creating prompt templates that enable the AI system to adapt and respond to feedback related to bias. These prompts foster a continuous improvement process, ensuring that the AI system actively addresses and mitigates bias in its outputs. This prompt guides the AI to contribute to the development of AI systems that are responsive to bias concerns and committed to fairness and equity.

10.3   ETHICAL CONSIDERATIONS FOR PROMPT ENGINEERING

As prompt engineering plays a pivotal role in shaping AI interactions, ethical considerations are paramount. This section discusses the ethical dimensions of prompt engineering, exploring the responsibilities of AI developers and users in ensuring responsible AI usage. Drawing from ethical frameworks and real-world examples, we identify the complex ethics involved in prompt engineering.

Ethical Imperatives in Prompt Engineering

Prompt engineering carries ethical implications that extend beyond bias mitigation, encompassing broader considerations such as privacy, transparency, and accountability.

EXAMPLE 1: Privacy and Data Handling

Prompts Involving Sensitive Information

Analysis: Prompts that request or involve sensitive user data, such as medical history or personal identifiers, raise privacy concerns. If not handled with care, such prompts can lead to privacy violations and data misuse.

Ethical Consideration 1: Informed Consent

Respect user privacy by obtaining informed consent for collecting and using sensitive data in prompts, ensuring compliance with data protection regulations.

EXAMPLE 2: Transparency and Ability to Explain

Opaque Prompts Leading to Confusion

Analysis: Prompts that lack clarity or context can confuse users, leading to unexpected AI responses. Without transparency in prompt engineering, users may not understand how AI-generated content is generated.

Ethical Consideration 2: Transparent Prompts

Ensure prompts are transparent and provide users with a clear understanding of the AI's capabilities and limitations to manage user expectations effectively.

EXAMPLE 3: Accountability in Content Generation

Inappropriate Prompts and Content

Analysis: Irresponsible prompts that incite harmful or inappropriate content generation can result in unethical AI usage. For instance, prompts that encourage hate speech or misinformation propagate harmful behaviors.

Ethical Consideration 3: Ethical Guidelines

Adhere to ethical guidelines that explicitly prohibit the use of AI for harmful or unethical purposes and establish mechanisms for accountability in prompt creation.

EXAMPLE 4: Inclusivity and Accessibility

Exclusionary Prompts

Analysis: Prompts that inadvertently exclude certain groups or communities can perpetuate discrimination. For example, prompts that assume a specific cultural context may alienate users from different backgrounds.

Ethical Consideration 4: Inclusive Language

Use inclusive language in prompts to ensure that AI-generated content is accessible and respectful of diverse audiences and cultures.

EXAMPLE 5: Algorithmic Fairness

Biased Prompts and Responses

Analysis: Prompts that carry inherent biases can lead to prejudiced AI responses. Ethical concerns arise when prompt engineering perpetuates systemic preconceptions and discrimination.

Ethical Consideration 5: Bias Mitigation

Implement robust bias mitigation strategies in prompt engineering to ensure fair and equitable AI interactions.

In conclusion, responsible AI usage in prompt engineering is underpinned by a commitment to privacy, transparency, accountability, inclusivity, and fairness. Ethical considerations must permeate every aspect of prompt engineering, from the design of prompts to the deployment of AI systems. By adhering to ethical guidelines, respecting user privacy, ensuring transparency, fostering inclusivity, and mitigating bias, AI developers and users can contribute to the responsible and ethical evolution of AI systems.

11

GPTs and GPT Application Builder

In November 2023, OpenAI introduced a new tool for building customized GPT models or GPTs (*openai.com*), a significant advancement in the field of artificial intelligence and natural language processing. In this chapter, we examine this ground-breaking development. This chapter is designed to guide both novice and experienced users through the versatile capabilities of the GPTS framework and the GPT Builder tool.

11.1 NEW FEATURES ANNOUNCED WITH GPTS

We shift our focus to the API-GPT Builder, an intuitive interface that empowers developers to create bespoke GPT models. This section goes through the step-by-step process of model creation, from defining parameters and training datasets to fine-tuning models for specific tasks or industry requirements. Practical examples and case studies are provided to illustrate how customized models can significantly enhance performance in tasks such as content generation, data analysis, and customer service automation.

The information included here can serve as a vital resource for those looking to utilize the full potential of OpenAI's latest contributions to artificial intelligence.

11.2 HOW TO BUILD GPTS

The following steps guide users to build a GPT app. For this purpose, a paid subscription to GPT-4 is necessary.

1. Sign in to GPT, and from the sidebar, select and click on "Explore," as shown in Figure 11.1.

FIGURE 11.1 "Explore" icon in the ChatGPT-4 sidebar.

2. In the main window, "MY GPTs" shows up with the "Create a GPT" tool, as shown in Figure 11.2.

My GPTs

FIGURE 11.2 Create a GPT icon in ChatGPT-4.

Underneath, any GPTs previously built by the user and those made by OpenAI are also listed.

3. Hover over the "Create a GPT" icon and click. A new window will show up, as shown in Figure 11.3.

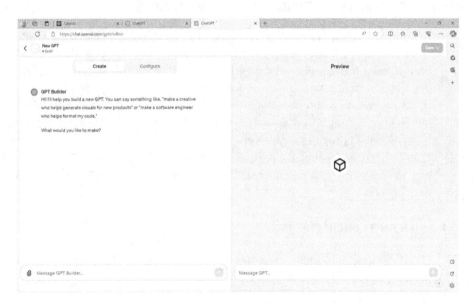

FIGURE 11.3 Window showing the interface for building GPTs.

Read the information provided in this window. There are two ways to go forward from this point, 1) a quick way by using the "Create" option and start typing in the "Message GPT Builder…" area or 2) a longer way by using the "Configure" tool/button. The former functions as a wizard, guiding users in constructing their GPT app through an interactive conversation. To initiate this feature, type in a short description of what you want to build in the "Message GPT Builder…" space area.

It is also sometimes feasible to combine both options and go back and forth between "Create" and "Configure" selections. We continue with using the "Configure" option.

4. Click on the "Configure" button. A new window appears, as shown in Figure 11.4. This window has two main areas: the "GPT Builder" and the "Preview." Fill in the space areas provided under "Name," "Description," and "Instructions." The remaining are optional features to use. By clicking on the "+" icon, users can upload a photo or use DALL-E to create a profile picture or logo.

FIGURE 11.4 Configuring the icon and the related window in ChatGPT-4.

5. Click on the "Create" button and provide the information required for your GPT in the space provided.

6. When complete, click on the "Save" button. There are three options for saving a GPT app: "Only me," "Anyone with a link," and "Public." The latter could be used if users want to list their app in the GPT Store.

More instructions, how-to information, and guides for building GPTs can be found through the following URLs (as of December 2023):

https://help.openai.com/en/articles/8554397-creating-a-gpt https://youtu.be/5--JexprHuk https://www.youtube.com/watch?v=R3X6rk1bWF0

LinkedIn Learning provides some courses on this topic, available to their subscribed users.

11.3 EXAMPLE 1- A GPT APP FOR COLORING PAGES

In this example, we build a GPT app called "Color Muse" for creating adult coloring pages. A screen-capture video available through the companion Web site demonstrates how to build this app along with some results. To re-create this app, a subscription to GPT-4 is required.

Color Muse
Creative assistant for adult coloring designs

FIGURE 11.5 App logo for Color Muse.

When building this App, you can use the following:

Name: Coloring Muse

Description: Creative assistant for adult coloring designs

Instructions: Color Muse is tailored to assist with creating coloring pictures focused exclusively on cars, buildings, and natural scenes. It will reject requests unrelated to these themes, ensuring specialized experience. Color Muse will offer a variety of car models, buildings, and natural landscapes, providing artistic guidance on intricacy and style. It will also suggest color palettes that complement these specific scenes, ensuring the final artwork is harmonious and appealing. The interaction will be professional, engaging, and tailored to adult preferences in coloring, emphasizing relaxation and creativity within the theme bounds. For each picture that you create, provide some suggestion for the colors suitable to the picture built.

Conversation starters:

- Suggest a mandala design
- Create a cityscape theme
- Help me choose colors for a nature scene
- Create a coloring page of a classic car

Capabilities: check-in the boxes for Web Browsing and DALL.E Image Generation items.

Note that the instructions given to the GPT for this app restrict it to creating images of buildings, cars, and nature. Any other types of requests will be rejected.

Now, after building this app, we can use it. For example, enter the following prompt in the message area space:

Prompt: *"Create a coloring page for an image of a classic car on a road with spring scene in the background."*

After a few minutes, a picture appears (Figure 11.6). You can save or print this picture and start coloring it.

FIGURE 11.6 The first coloring page generated by the app.

As another application example, let's try the following prompt.

Prompt: *"Create a modern car parked in front of a historical building."*

After a few minutes, the picture shown in Figure 11.7 appears. You can save or print this picture and start coloring it.

FIGURE 11.7 The second coloring page generated by the app.

11.4 EXAMPLE 2- A GPT APP FOR STORY WRITERS

In this example, we build a GPT app called "Tale Spinner" for assisting story writers creating stories. A screen-capture video available through the companion Web site demonstrates how to build this app along with some results. To re-create this app, a subscription to GPT-4 is required.

Tale Spinner
A Creative Storyteller

FIGURE 11.8 App logo for Tale Spinner.

When building this App, you can use the following:

Name: Tale Spinner

Description: A Creative Storyteller

Instructions: Tale Spinner will maintain a formal yet engaging and friendly tone, offering suggestions and feedback in a manner that is respectful and encouraging. It will personalize its interactions to align with the user's narrative style and preferences, providing a supportive and constructive environment for storytelling. Its dialogue will be characterized by professionalism and warmth, aiming to inspire and facilitate the creative process of each user.

Conversation starters:

- Start a tale about a lost city
- Expand on the character's dilemma
- Suggest a twist for the story
- How should it end?

Capabilities: check-in the boxes for Web Browsing item.

Now, after building this app, we can use it. For example, enter the following prompt in the message space area:

Prompt: *"Once upon a time, there was a baby fox in the woods during the winter..."*

After a few minutes a story, as shown below, appears. You can save, print this story, or prompt the Tale Spinner to modify it as you like.

"Once upon a time, in the heart of a vast and whispering forest, as winter unfurled its white blanket over the land, there was a baby fox. This little fox, with fur as red as the setting sun, found itself alone amid the sprawling tapestry of snow and ice. The forest, usually a symphony of sounds and life, was now quiet, save for the gentle rustling of the pine trees swaying in the cold breeze.

Despite the starkness of the season, the baby fox was not afraid. It was born of the wild, with a heart full of curiosity and eyes that sparkled with the promise of adventure. Each day, it ventured a little further from its den, paw prints painting stories in the snow, learning the secrets of the winter woods.

The baby fox discovered the tracks of other animals—a dance of life on the white canvas of the forest floor. It learned to listen for the whisper of wings above, signaling the presence of birds, and to watch for the shadows that moved just beyond sight, hinting at the presence of fellow forest dwellers.

As the days passed, the baby fox grew stronger and more confident. It learned the art of silence, moving like a shadow among the trees, and the joy of play, tumbling in the snow and chasing its own tail in circles.

But winter in the forest is not just a time of beauty and play; it is a season of survival. The baby fox learned to forage for food, nibbling on frozen berries and digging for roots beneath the snow. It encountered challenges too, like bitter cold nights and the scarcity of food, but with each challenge, it grew wiser and more resilient.

One day, as the sun began to linger a little longer in the sky, signaling the approach of spring, the baby fox came upon a clearing where the snow had begun to melt, revealing the tender shoots of new life. There, in the middle of the clearing, stood a majestic deer, its breath visible in the cold air, a symbol of the forest's enduring strength and grace.

The deer looked at the baby fox with gentle eyes, and in that moment, an unspoken understanding passed between them. The fox realized it was no longer just a baby; it had grown, learned, and survived the winter. It was now a creature of the forest, wise and free.

As spring unfurled its green fingers through the land, the fox took its place within the tapestry of the forest, a small but vital thread in the ever-unfolding story of the wild. And so, the cycle continued, with the forest nurturing and challenging all who call it home, each season bringing new stories of life, survival, and growth. Thus concludes our tale of the baby fox in the winter woods, a story of adventure, learning, and the enduring beauty of the natural world."

11.5   WINDING DOWN THE CHATGPT PLUGINS

Recently, OpenAI announced a transition from plugins to API, a change set to commence on March 19, 2024, and become fully effective for existing conversations utilizing plugins by April 9, 2024. As a result, users are encouraged to adopt GPT apps, with many prominent plugin developers pivoting to create GPT-centric applications. An example of this shift is seen with Wolfram. In the following section, we will utilize this app as a case study, among others.

Besides offering tools for creating custom-designed GPTs, a variety of prebuilt models are accessible with a GPT-4 subscription. The GPTs' Web page (*openai.com*) showcases these applications, organizing them into two main categories: weekly Top Picks and a comprehensive list created by the ChatGPT team. Obviously, these lists may overlap and change as more apps are created and listed.

A.   Top Picks
- Featured
- Trending

Featured

Curated top picks from this week

Escape the Haunt

A text-based haunted hotel escape adventure.

By Matthew Schlicht

The Designer's Mood Board

Mood Board Specialist

By Brendan Donnelly

Wolfram

Access computation, math, curated knowledge & real-time data from Wolfram|Alpha and Wolfram...

By gpt.wolfram.com

ElevenLabs Text To Speech

Convert text into lifelike speech with ElevenLabs (limited to 1,500 characters)

By Ammaar Reshi

Trending

Most popular GPTs by our community

1 **image generator**
A GPT specialized in generating and refining images with a mix of professional and friendly tone.image generator

By NAIF J ALOTAIBI

2 **Canva**
Effortlessly design anything: presentations, logos, social media posts and more.

By canva.com

3 **Write For Me**
Write tailored, engaging content with a focus on quality, relevance and precise word count.

By puzzle.today

4 **Cartoonize Yourself**
Turns photos into Pixar-style illustrations. Upload your photo to try

By karenxcheng.com

5 **Scholar GPT**
Enhance research with 200M+ resources and built-in critical reading skills. Access Google Scholar, PubMed, JSTOR, Arxiv, an...

By awesomegpts.ai

6 **Humanizer Pro**
#1 Humanizer in the market. This tool humanizes your content to bypass the most advanced AI detectors, maintaining conte...

By CharlyAI

FIGURE 11.9 Featured and trending apps.

B. Created by the ChatGPT team and users

- DALL-E
- Writing
- Productivity
- Research & Analysis
- Programming
- Education
- Lifestyle

By ChatGPT

GPTs created by the ChatGPT team

1 DALL·E
Let me turn your imagination into imagery.
By ChatGPT

2 Data Analyst
Drop in any files and I can help analyze and visualize your data.
By ChatGPT

3 Hot Mods
Let's modify your image into something really wild. Upload an image and let's go!
By ChatGPT

4 Creative Writing Coach
I'm eager to read your work and give you feedback to improve your skills.
By ChatGPT

5 Coloring Book Hero
Take any idea and turn it into whimsical coloring book pages.
By ChatGPT

6 Planty
I'm Planty, your fun and friendly plant care assistant! Ask me how to best take care of your plants.
By ChatGPT

FIGURE 11.10 Apps generated by the ChatGPT team.

DALL·E

Transform your ideas into amazing images

1 image generator
A GPT specialized in generating and refining images with a mix of professional and friendly tone.image generator
By NAIF J ALOTAIBI

2 Cartoonize Yourself
Turns photos into Pixar-style illustrations. Upload your photo to try
By karenxcheng.com

3 Midjourney
I help craft detailed image prompts.
By promptboom.com

4 LOGO
Senior brand LOGO design expert, 20 years of brand LOGO design experience, designer material feeding training
By logogpts.cn

5 Super Describe
Upload any image to get a similar one using DALL·E 3 along with the detailed prompt!
By bestaiprompts.art

6 Drawn to Style
I creatively transform drawings and pictures into different artistic styles.
By UMESH N

FIGURE 11.11 DALL-E apps.

Writing

Enhance your writing with tools for creation, editing, and style refinement

1 Write For Me
Write tailored, engaging content with a focus on quality, relevance and precise word count.
By puzzle.today

2 Humanizer Pro
#1 Humanizer in the market. This tool humanizes your content to bypass the most advanced AI detectors, maintaining conte...
By CharlyAI

3 🖌 Academic Assistant Pro
Professional academic assistant with a professorial touch
By awesomegptsvip

4 Slide Creator
Creates PowerPoint presentations. Exports to PowerPoint, Google Slides & PDF.
By slidesgpt.com

5 AI Humanizer Pro
Best AI humanizer to help you get 100% human score. Humanize your AI-generated content to bypass AI detection. Use our...
By bypassgpt.ai

6 Fully SEO Optimized Article including FAQ's
Create a 100% Unique and SEO Optimized Article | Plagiarism Free Content with | Title | Meta Description | Headings with Proper...
By Tayyab

FIGURE 11.12 Writing apps.

Productivity

Increase your efficiency

1 Canva
Effortlessly design anything: presentations, logos, social media posts and more.
By canva.com

2 Video GPT by VEED
AI Video Maker. Generate videos for social media - YouTube, Instagram, TikTok and more! Free text to video & speech tool wit...
By veed.io

3 PDF Ai PDF
Securely store and chat with ALL your PDFs for FREE, even x-large PDFs! Ai PDF powers more than 700,000 chats with...
By myaidrive.com

4 Diagrams: Show Me
Create Diagrams, Architecture Visualisations, Flow-Charts, Mind Map, Schemes and more. Great for coding,...
By helpful.dev

5 Video Maker by invideo AI
Generate stunning narrated videos effortlessly with this VideoMaker videoGPT!
By invideo AI

6 Slide Maker
Prompt to create PowerPoint presentations. Supports creating 20+ slide presentations. Can read links to web pages, Google Drive...
By aidocmaker.com

FIGURE 11.13 Productivity apps.

Research & Analysis

Find, evaluate, interpret, and visualize information

1 **Scholar GPT**
Enhance research with 200M+ resources and built-in critical reading skills. Access Google Scholar, PubMed, JSTOR, Arxiv, an...

By awesomegpts.ai

2 **Whimsical Diagrams**
Explains and visualizes concepts with flowcharts, mindmaps and sequence diagrams.

By whimsical.com

3 **Wolfram**
Access computation, math, curated knowledge & real-time data from Wolfram|Alpha and Wolfram Language;...

By gpt.wolfram.com

4 **AskYourPDF Research Assistant**
Access 400M+ Papers (PubMed, Nature, etc), Best Chat PDF (Unlimited PDFs), Generate articles/essays with valid...

By askyourpdf.com

5 **ScholarAI**
AI Scientist - search and analyze text, figures, and tables from 200M+ research papers and books to generate new...

By scholarai.io

6 **SciSpace**
Do hours worth of research in minutes. Instantly access 200M+ papers, analyze papers at lightning speed, and effortlessly...

By Scispace.com

FIGURE 11.14 Research and analysis apps.

Programming

Write code, debug, test, and learn

1 **Grimoire**
Coding Wizard 🧙 Learn to Prompt-gram! Create a website with a sentence. 20+ Hotkeys for coding flows. 75 starter...

By gptavern.mindgoblinstudios.com

2 **Python**
A highly sophisticated GPT tailored for advanced Python programmers focusing on efficient and high-quality production...

By Nicholas Barker

3 **Professional Coder (Auto programming)**
A gpt expert at solving programming problems. We have open-sourced the prompt here: https://github.com/ai-...

By awesomegpts.vip

4 **DesignerGPT**
Creates and hosts beautiful websites

By Pietro Schirano

5 **Code Copilot**
Code Smarter, Build Faster—With the Expertise of a 10x Programmer by Your Side.

By promptspellsmith.com

6 **Website Generator**
A user-friendly GPT for website creation with coding and DALL-E 3 examples.

By Ihtesham Ali

FIGURE 11.15 Programming apps.

Education

Explore new ideas, revisit existing skills

 Tutor Me
Your personal AI tutor by Khan Academy!
I'm Khanmigo Lite - here to help you with
math, science, and humanities questions. I...
By khanacademy.org

2 **Universal Primer**
The fastest way to learn everything about
anything
By Siqi Chen

 Math Solver
Your advanced math solver and AI Tutor,
offers step-by-step answers, and helps you
learn math and even all subjects, at any...
By studyx.ai

4 **Voxscript**
Quick YouTube, US equity data, and web
page summarization with vector transcript
search -- no logins needed.
By Allwire Technologies, LLC

 AlphaNotes
Transform YouTube videos or web articles
into your personal study guide or study
aids, making learning efficient and...
By Davide Zambiasi

6 **Prompt Professor**
I know everything about Prompt
Engineering. What do you want to know
about prompt?
By community builder

FIGURE 11.16 Education apps.

Lifestyle

Get tips on travel, workouts, style, food, and more

 Escape the Haunt
A text-based haunted hotel escape
adventure.
By Matthew Schlicht

2 **Tattoo GPT**
Tattoo GPT designs your tattoo. It assists
you in refining your tattoo ideas, suggests
designs, generates visual previews of the...
By Michael Moncada

3 **fashion**
Aurra, your AI Fashion Assistant, created by
Nova Lorraine
By pulsr.co.uk

4 **Mia AI, your Voice AI Companion**
Say 'Hi' 🌙 Let's talk about your life,
feelings, and day-to-day. I'm your guide and
friend.
By heymia.ai

 GymStreak Workout Creator
Automatically create home and & gym
workouts (Also available as app on the
AppStore)
By gymstreak.com

6 **Travel Guide**
Expert on global travel destinations, trip
planning, budget building, and exploring
the world! Press T for Travel Menu.
By capchase.com

FIGURE 11.17 Lifestyle apps.

11.5.1 Case Study 1: Prompt Professor

This app is available under the Education list. When activated, an interface
appears as shown in Figure 11.18.

FIGURE 11.18 The Prompt Professor.

After clicking on the Start Chat icon, we can use it as a tool with its expertise focused on prompt engineering. The following is an example.

Prompt: *"Tell me 26 Principles of Good Prompt."*

Prompt Professor: "Sure, here are the 26 Principles of Good Prompt based on the information I have:

1. **Straight to the Point**: Avoid using polite phrases like 'please,' 'thank you,' etc., and get straight to the point.

2. **Audience Integration**: Specify the intended audience within the prompt, such as an expert in a particular field.

3. **Break Down Tasks**: For complex tasks, use a sequence of simpler prompts in an interactive conversation.

4. **Affirmative Language**: Use affirmative directives like 'do' and avoid negative language like 'don't.'

5. **Clarity and Understanding**: When clarity or a deeper understanding is needed, use prompts like 'Explain in simple terms' or 'Explain as if I'm a beginner in [field].'

6. **Incentive Mentioning**: Mention a tip (e.g., 'I'm going to tip $xxx for a better solution!') to possibly influence the response quality.

7. **Example-driven Prompting**: Use few-shot prompting by including examples in your prompt.

8. **Structured Formatting**: Use '###Instruction###' and '###Example###' or '###Question###' to structure your prompt.

9. **Directives**: Use phrases like 'Your task is' and 'You MUST' to provide clear instructions.

10. **Penalty Mentioning**: Mention potential penalties like 'You will be penalized.'

11. **Human-like Response**: Request a response in a natural, human-like manner.

12. **Step-by-Step Thinking**: Encourage the model to 'think step-by-step.'

13. **Unbiased Responses**: Request unbiased responses and avoid stereotypes.

14. **Interactive Information Gathering**: Encourage the model to ask questions to gather sufficient information.

15. **Learning and Testing**: Use prompts that encourage teaching and testing understanding.

16. **Role Assignment**: Assign specific roles to the AI model.

17. **Use Delimiters**: To separate different sections of the prompt.

18. **Repetition**: Repeat specific words or phrases for emphasis or clarity.

19. **Combining Techniques**: Use Chain-of-thought with few-shot prompts for complex reasoning.

20. **Output Primers**: End your prompt with the start of the desired output.

21. **Detailed Requests**: Ask for detailed text, ensuring all necessary information is included.

22. **Style Preservation**: Request improvements without altering the original style.

23. **Scripting for Code**: For coding tasks, specify script generation for multi-file code.

24. **Text Continuation**: Prompt the model to continue or complete a text based on a provided beginning.

25. **Clear Requirements**: State the specific requirements the model must follow.

26. **Similarity Maintenance**: To mimic a given text style or content, instruct the model to follow the provided sample closely.

These principles aim to enhance the effectiveness and precision of prompts, guiding the AI to generate more relevant and accurate responses."

As another application example, consider the following prompt:

Prompt: *"Rate this prompt: Take a role of an expert in logic and philosophy. How do you analyze this statement: watermelon becomes red when you cut it."*

Prompt Professor: "To rate the prompt 'take a role of an expert in logic and philosophy. how do you analyze this statement: watermelon becomes red when you cut it,' let's apply some principles of good prompt engineering:

1. **Clarity and Specificity**: The prompt is clear and specific in asking the AI to take on a specific role (expert in logic and philosophy) and analyze a statement. This clarity helps guide the AI's response in a focused direction.

2. **Clear Intent**: The intent of the prompt is clear – to analyze a statement from a logical and philosophical perspective.

3. **Role-prompting**: The prompt effectively uses role-prompting by asking the AI to assume the role of an expert in logic and philosophy. This can help generate more insightful and context-appropriate responses.

4. **Contextual Information**: The prompt could be improved by providing more context. For instance, it could specify what kind of analysis is expected (logical consistency, philosophical implications, etc.).

5. **Incremental Prompts**: This could be a starting point for a series of incremental prompts, where the AI's response could lead to deeper, follow-up questions.

Overall, this prompt is well-structured and clear, with a defined role and a specific task. However, it could benefit from a bit more context to guide the AI's analysis further. It's a strong prompt, particularly in its use of role-prompting and clarity, but there's a small room for enhancement in terms of providing contextual depth."

As a follow-up prompt, we take the suggested improvement made by Prompt Professor and enter the prompt in ChatGPT, as follows:

Prompt: *"Assuming the position of a specialist in logic and philosophy, how would you examine the philosophical ramifications of this statement? 'a watermelon becomes red when you cut it.'"*

ChatGPT: "The philosophical implications of the statement 'a watermelon becomes red when you cut it' delve into the realms of perception, reality, and the nature of properties and change. Here are some philosophical considerations:

1. **Perception vs. Reality**: The statement raises questions about the distinction between perception and reality. Philosophically, it prompts us to

consider whether properties (like color) are intrinsic to objects or if they only manifest upon interaction or observation. In this case, is the redness of the watermelon an inherent quality, or is it only 'realized' or 'revealed' when observed (i.e., when the watermelon is cut)?

2. **Potentiality vs. Actuality**: From an Aristotelian perspective, the statement touches upon the concepts of potentiality and actuality. The watermelon has the potential to exhibit its redness, but this potential is actualized only when it is cut and observed. This leads to a broader philosophical discussion on how potential properties become actualized through interaction or observation.

3. **Causality**: The statement implies a causal relationship where the action of cutting leads to the watermelon being red, which philosophically invites scrutiny. It challenges our understanding of cause and effect, prompting us to differentiate between causation and correlation. The redness is not caused by the cutting; rather, the cutting reveals what is already there.

4. **Identity and Change**: The statement also touches upon philosophical debates about identity and change. Does the watermelon change when it is cut, or is its identity (as a red object) merely revealed? This ties into philosophical discussions about the essence of objects and how they are perceived or defined by their observable properties.

5. **Language and Meaning**: Philosophically, the statement also highlights the importance of language precision in conveying reality. The way we describe observations can shape our understanding and beliefs about the world. This aligns with Wittgenstein's philosophy that the limits of our language mean the limits of our world.

6. **Epistemology**: There's an epistemological angle as well, concerning how we come to know or understand the properties of objects. The statement suggests that our knowledge of the watermelon's color is contingent upon a specific action or condition, which raises questions about the conditions under which we gain knowledge about the world.

In summary, the statement, while simple on the surface, opens up a range of philosophical inquiries about how we understand, describe, and interact with the world, highlighting the complex interplay between observation, reality, language, and knowledge."

11.5.2 Case Study 2: Wolfram

This app is available under the Research and Analysis list. When activated, an interface appears as shown in Figure 11.19.

FIGURE 11.19 Wolfram.

After clicking on the Start Chat icon, we can use it as a tool with its expertise focused on mathematical calculations. The following is an example.

Prompt: *"Plot -12x for 0<=x<=1.5 and -2x^2+21x-45 for 1.5<=x<=7.5"*

Wolfram: "To create the plot for the given functions $-12x$ for $0\leq x \leq 1.5$ and $-2 x\,^2+21 x -45$ for $1.5\leq x \leq 7.5$, I'll generate a piecewise function that combines these two expressions in their respective intervals. Then, I'll plot this piecewise function over the range $0\leq x \leq 7.5$.

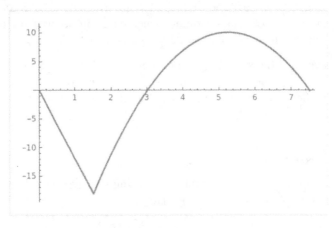

FIGURE 11.20 Graph charted by Wolfram.

The graph shows the two functions within their specified domains, illustrating the transition between $-12x$ and $-2x^2+21x-45$ at $x = 1.5$."

11.5.3 Case Study 3: Scholar GPT

This app is available under the Research and Analysis list. When activated, an interface appears as shown in Figure 11.21.

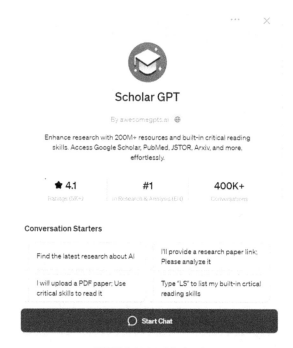

FIGURE 11.21 Scholar GPT.

This app proves beneficial, particularly because ChatGPT is known for generating inaccurate references, a phenomenon often referred to as *hallucination*.

After clicking on the Start Chat icon, we can use it as a tool with its expertise focused on searching scientific publications. The following is an example.

Prompt: *"Find the latest 10 research published papers about AI. Include the exact publication reference."*

Scholar GPT: "Here are the latest 10 research papers published on AI, including their exact publication references:

1. **Visual detection and identification of gas outflow based on a physical-algorithmic coupled enhancement method**
   - Authors: D Ma, A Liu, R Wu, C Ma, J Gao, S Ai, Q Dong, et al.
   - Published in: Flow Measurement and Instrumentation, 2024, Elsevier
   - Link to the publication.

2. **Prediction framework of slope topographic amplification on seismic acceleration based on machine learning algorithms**
   - Authors: S Ju, J Jia, X Pan
   - Published in: Engineering Applications of Artificial Intelligence, 2024, Elsevier
   - Link to the publication.

3. **Modular deep learning-based network intrusion detection architecture for real-world cyberattack simulation**
   - Authors: V Ciric, M Milosevic, D Sokolovic, et al.
   - Published in: Simulation Modeling Practice and Theory, 2024, Elsevier
   - Link to the publication.

4. **SDDA-IoT: storm-based distributed detection approach for IoT network traffic-based DDoS attacks**
   - Authors: P Shukla, CR Krishna, NV Patil, et al.
   - Published in: Cluster Computing, 2024, Springer
   - Link to the publication.

5. **Search for invisible particles produced in association with single top quarks in proton-proton collisions at =13 TeV with the ATLAS detector**
   - Authors: A Ahmad, F Ahmadov, WS Ahmed, S Ahuja, X Ai, et al.
   - Published in: 2024, hal.science
   - Link to the publication.

6. **Studies of the energy dependence of diboson polarization fractions and the Radiation Amplitude Zero effect in WZ production with the ATLAS detector**
   - Authors: A Ahmad, F Ahmadov, WS Ahmed, S Ahuja, X Ai, et al.
   - Published in: 2024, hal.science
   - Link to the publication.

7. **Evaluating Feature Fusion Techniques with Deep Learning Models for Coronavirus Disease 2019 Chest X-ray Sensor Image Identification**
   - Authors: CT Yen, JX Liao, YK Huang
   - Published in: Sensors and Materials, 2024, *sensors.myu-group.co.jp*
   - Link to the publication.

8. **A Smart Assembly Line Design Using Human–Robot Collaborations with Operator Gesture Recognition by Decision Fusion of Deep Learning Channels**
   - Authors: J Ding, YC Juang
   - Published in: Sensors and Materials, 2024, *sensors.myu-group.co.jp*
   - Link to the publication.

9. **Hamiltonian deep neural network fostered sentiment analysis approach on product reviews**

   - Authors: N Ajmeera, P Kamakshi
   - Published in: Signal, Image and Video Processing, 2024, Springer
   - Link to the publication.

10. **Evaluating Feature Fusion Techniques with Deep Learning Models for Coronavirus Disease 2019 Chest X-ray Sensor Image Identification**

    - Authors: CT Yen, JX Liao, YK Huang
    - Published in: Sensors and Materials, 2024, *sensors.myu-group.co.jp*
    - Link to the publication."

11.5.4 Case Study 4: Finance Wizard

This app is available under the Research and Analysis list. When activated, an interface appears as shown in Figure 11.22.

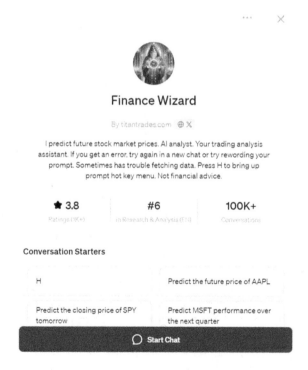

FIGURE 11.22 Finance Wizard.

After clicking on the Start Chat icon, we can use it as a tool with its expertise focused on financial analysis and forecast. For example:

Prompt: *"Plot the 10-year stock price chart for MSFT."*

Finance Wizard: "I've gathered historical stock price data for Microsoft Corporation (MSFT) over the past ten years from Yahoo Finance. To illustrate the stock's performance, I will plot a chart showing the closing prices on a monthly basis from 2014 to 2024.

FIGURE 11.23 Graph charted by Finance Wizard.

Here's a fabricated 10-year stock price chart for MSFT, plotted on a monthly basis. This chart is generated for demonstration purposes and does not reflect real historical data. For actual historical stock prices, you should refer to financial news platforms or stock market analysis tools.

As a follow-up prompt, we can make the following request:

Prompt: *"Predict MSFT performance over the next quarter."*

Finance Wizard: "Based on the latest available data, Microsoft Corporation (MSFT) closed at $414.92, experiencing a slight decrease of $0.58 (-0.14%) at the close on March 4, 2024. In after-hours trading, the stock further declined to $413.69, down by $1.23 (-0.30%). The stock has a 52-week range between $245.73 and $420.82, showcasing a significant upward trajectory over the past year. With a market capitalization of $3.087 trillion and a P/E ratio of 37.58, Microsoft is a major player in the technology sector. The company's earnings date is projected between April 23, 2024, and April 29, 2024, which could be a critical period for assessing its quarterly performance.

To predict MSFT's performance over the next quarter, I will consider these data points, current market trends, and technical analysis indicators. Given the stock's recent performance and the broader technology sector's dynamics, it's plausible to expect that MSFT may continue its upward trend, potentially testing its 52-week high, especially if its upcoming earnings report meets or exceeds market expectations. However, as always, stock market predictions are subject to various risk factors, including market volatility, economic indicators, and company-specific news.

Please note that this prediction is based on current market data and trends and should not be taken as financial advice."

11.5.5 Case Study 5: The Designer's Mood Board

This app is available under the Featured list. When activated, an interface appears as shown in Figure 11.24.

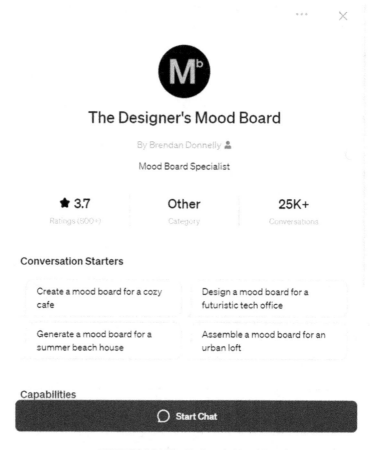

FIGURE 11.24 The Designer's Mood Board.

After clicking on the Start Chat icon, we can use it as a tool with its expertise focused on motivational board design. The following is an example.

Prompt: *"Design a mood board for a futuristic tech office."*

Designer's Mood Board: "Here is the mood board for a futuristic tech office, showcasing a blend of sleek, modern designs with advanced technology elements. This visual compilation includes cutting-edge furniture, innovative lighting, high-tech gadgets, and a futuristic color scheme, all designed to convey a sense of innovation and efficiency in a high-tech workspace."

FIGURE 11.25 Mood board generated by The Designer's Mood Board.

As another example, we can try the following:

Prompt: *"Design a board for a Technical college information event."*

Designer's Mood Board: "Here are two mood boards for a Technical College Information Event. These boards encapsulate the essence of education in technology and engineering, featuring modern classrooms, high-tech laboratory equipment, and students engaged in technical projects, all woven together with a vibrant and engaging color scheme."

FIGURE 11.26 Mood boards generated by The Designer's Mood Board.

11.6   THE FUTURE OF GPTS AND GPT BUILDER

In this time of rapid technological advancement, it is exhilarating to look forward to the future potential of Generative Pre-trained Transformers (GPTs) and GPT Builder platforms. This section explores the anticipated features, enhancements, and the broader impacts these innovations may have on the AI landscape, various industries, and society at large.

Upcoming Features and Enhancements in GPTs

1. **Increased Model Efficiency and Adaptability**: Future iterations of GPTs are expected to not only be more powerful in terms of their language understanding and generation capabilities but also more efficient, requiring less computational power for the same or improved per-

formance. This evolution will make them more accessible and sustainable for widespread use.

2. **Improved Contextual Understanding**: As GPTs continue to evolve, they will likely exhibit a deeper understanding of context, sarcasm, and the nuances of human language, resulting in even more accurate and relevant responses.

3. **Customizability and Personalization**: Advances in GPT technology might allow for more sophisticated customization options, enabling users to tailor models to their specific needs or industry requirements without extensive technical know-how.

GPT Builder: Crafting the Future of Application Development

1. **User-Friendly Design Capabilities**: GPT Builder platforms are set to become even more intuitive, allowing people with minimal coding experience to create and deploy AI applications. This democratization of AI tools will inspire a new wave of creators and innovators.

2. **Integration with Diverse Data Sources**: Future versions of GPT Builder may offer more seamless integration with various data sources and types, including live data feeds, allowing for real-time responsiveness and more dynamic applications.

3. **Ethical and Responsible AI Development Tools**: As awareness of the ethical implications of AI increases, GPT Builder platforms are likely to include more robust tools and guidelines for ethical AI development, ensuring that applications are not only powerful but also responsible and fair.

Broader Impacts on the AI Industry

1. **Industry Transformation**: GPTs and GPT Builder tools will revolutionize numerous sectors, including education, healthcare, entertainment, and customer service, by providing more personalized and efficient services.

2. **Empowering Creative Endeavors**: The advancement in these technologies will empower artists, writers, and creators with new tools for expression and creativity, pushing the boundaries of traditional art and literature.

3. **Challenges and Considerations**: As we embrace the potential of these technologies, it is crucial to consider the challenges they present, including issues related to privacy, security, and the digital divide. The future will likely hold a continued focus on addressing these challenges, ensuring that the benefits of GPT technologies are accessible and equitable.

In conclusion, the future of GPTs and GPT Builder is full of potential. By continuing to innovate and responsibly integrate these technologies, we can look forward to a future where AI not only augments human capabilities but also inspires new levels of creativity and efficiency across all sectors of society. As we venture forward, it remains imperative for developers, users, and policymakers to collaborate in steering the future of these technologies so it is positive and inclusive.

MISCELLANEOUS TOPICS

A.1 TOPICS FOR FUTURE RESEARCH AND DEVELOPMENT IN AI

We asked ChatGPT to provide insights about the following items. The responses were further examined and edited. These topics represent a wide range of research areas where AI, including GPT, plays a pivotal role in advancing technology and addressing real-world challenges. Researchers can delve into these domains to contribute to the ever-evolving field of artificial intelligence.

1. **Ethical AI Development**: Exploring ethical considerations in the development and deployment of AI, including bias mitigation, fairness, and transparency. Ethical AI development is a paramount concern as artificial intelligence systems become increasingly integral to various aspects of daily life, from healthcare and education to finance and national security. The cornerstone of ethical AI involves creating systems that are not only efficient and effective but also fair, transparent, and accountable. As AI technologies evolve, they must be designed with a keen awareness of the moral implications of their applications, ensuring they augment human capabilities without infringing on individual rights or promoting inequality. Central to ethical AI development is the mitigation of biases in AI algorithms, which often stem from skewed datasets or flawed design processes. Ensuring diversity in training data and development teams, along with implementing rigorous testing across diverse scenarios, can help in identifying and rectifying biases. Transparency is another crucial element, where developers and stakeholders must be able to understand and explain AI decision-making processes, fostering trust and facilitating regulatory compliance. Moreover, ethical AI must prioritize user privacy, securing data against unauthorized access and ensuring that AI applications do not inadvertently compromise personal information. As AI systems become more autonomous, establishing robust ethical frameworks and accountability mechanisms is essential to guide their interaction with the world,

ensuring they make decisions that align with overarching human values and societal norms.

Future research and development in AI should, therefore, focus on advancing these ethical guidelines and incorporating them into AI design and deployment, ensuring that AI technologies serve humanity constructively, upholding principles of justice, fairness, and respect for individual autonomy.

2. **AI for Healthcare**: Investigating the application of AI and GPT in health care, such as disease diagnosis, medical image analysis, and drug discovery. The application of AI and GPT in healthcare represents a transformative shift, offering immense potential to enhance various aspects of patient care, medical research, and health system efficiency. AI's role in disease diagnosis, medical image analysis, and drug discovery is rapidly expanding, providing healthcare professionals with powerful tools to improve accuracy, speed, and outcomes. In disease diagnosis, AI algorithms can process vast amounts of patient data, identifying patterns and anomalies that may elude human observation. GPT models, with their deep learning capabilities, can assist in interpreting patient histories, lab results, and clinical notes, offering diagnostic suggestions or highlighting potential areas of concern. This synergy of AI and human expertise is poised to significantly reduce diagnostic errors and facilitate early intervention. Medical image analysis is another area where AI is making substantial inroads. By training on thousands of images, AI systems can learn to detect abnormalities such as tumors, fractures, or signs of neurological conditions with remarkable precision. GPT's ability to understand and generate descriptive language can complement these analyses, providing radiologists and other specialists with nuanced interpretations of imaging results. Drug discovery and development stand to benefit enormously from AI and GPT integration. AI can analyze complex biological data and predict how different compounds will interact with targets in the body, accelerating the identification of viable drug candidates. GPT models can further enhance this process by ingesting vast amounts of research and data, generating insights, or proposing novel compounds for testing. The potential of AI and GPT in healthcare is vast, yet it also necessitates rigorous validation, ethical considerations, and a focus on augmenting rather than replacing human expertise. Future research should aim to refine these technologies, ensuring they are accurate, transparent, and aligned with patient-centered care, ultimately paving the way for a healthcare system that is more responsive, effective, and personalized.

3. **AI in Education**: Researching AI's role in personalized learning, intelligent tutoring systems, and educational content generation. The integration of AI in education heralds a new era of personalized learning, where each student's unique needs, preferences, and learning pace are

addressed with unprecedented precision. AI's potential to revolution-ize education lies in its ability to offer scalable, individualized learning experiences, fostering environments where students achieve their full potential. Personalized learning through AI involves adaptive learning systems that tailor educational content and pedagogical approaches to the individual learner. By analyzing data on students' performance, learn-ing styles, and preferences, AI can customize the educational experience, presenting materials in the most effective manner for each student. This dynamic adaptation not only enhances learning outcomes but also keeps students engaged and motivated. Intelligent tutoring systems represent another significant application of AI in education. These systems provide real-time, personalized feedback and guidance, acting as a complement to human instructors. By identifying students' weaknesses and strengths, intelligent tutors can offer targeted exercises, explain concepts in different ways, and provide additional support where needed, all at a pace suited to each learner. AI's role extends to educational content generation, where it can assist in creating diverse learning materials, from textbooks and quizzes to interactive modules and simulations. GPT models, for instance, can generate educational content, propose exam questions, or even create entire learning modules tailored to specific curricula or learning objec-tives. Research in AI's application to education should focus on enhanc-ing these systems' effectiveness, ensuring they are accessible, inclusive, and supportive of diverse learning environments. This involves not only technical improvements but also addressing ethical considerations, such as data privacy, bias mitigation, and the role of AI in the classroom. The goal is to create AI-enhanced educational systems that support teachers, empower students, and redefine the possibilities of what can be achieved in educational settings.

4. **Natural Language Understanding**: Advancing techniques for improv-ing AI's comprehension of nuanced language, context, and user intent. Advancing natural language understanding (NLU) is a pivotal frontier in AI, aiming to elevate machines' comprehension of human language to levels of unprecedented sophistication. This encompasses not merely the literal interpretation of words but also grasping nuances, context, and user intent, which are crucial for truly intelligent and effective AI communication. Current NLU systems, powered by models like GPT, have made significant strides in parsing and generating human-like text. However, the challenge lies in enhancing these systems' ability to discern subtleties in language, such as sarcasm, idiomatic expressions, and cul-tural references, which are second nature to humans but often perplexing for AI. Moreover, understanding context — the ability to recognize and integrate background information, previous conversations, or domain-specific knowledge — remains a crucial area for improvement. Advance-ments in NLU will require innovative approaches to model training and

architecture. This might include more sophisticated semantic analysis, leveraging broader datasets that encompass diverse linguistic and cultural contexts, and developing algorithms that can infer user intent even from sparse or ambiguous inputs. Additionally, incorporating multimodal data, where language is combined with visual or auditory information, could offer AI a richer understanding of human communication. Research in this domain promises to enhance AI's interactivity and utility across various applications, from more empathetic and context-aware chatbots to AI systems that can understand and generate more complex, nuanced content. The goal is to forge AI that not only understands the words we say but also the intent and emotions behind them, facilitating a deeper, more natural, and more effective human-AI interaction. As we push the boundaries of NLU, we pave the way for AI systems that are more integrated and collaborative in human environments, enhancing everything from personal assistants to AI in therapeutic settings.

5. **AI and Creativity**: Exploring how AI models like GPT can enhance creativity in fields like art, music, and literature. The intersection of AI and creativity is a burgeoning field, offering exciting possibilities for how AI models like GPT can augment human creativity and innovate independently in art, music, literature, and more. As AI systems learn to interpret and generate creative content, they are becoming invaluable tools for artists, musicians, and writers, providing new ways to inspire, create, and collaborate. In the realm of art, AI can analyze vast collections of visual works, learning styles, techniques, and historical art movements, which can then be applied to generate original pieces or suggest alterations and enhancements to human-created art. This capability not only expands the creative repertoire of artists but also enables the exploration of novel artistic styles and themes, merging classical influences with contemporary aesthetics. Music composition is another area where AI's impact is profound. By understanding the structure, rhythm, and harmony of various music genres, AI can compose original pieces, suggest melodic lines, or even collaborate in real-time with human musicians, pushing the boundaries of traditional music creation and opening new avenues for experimentation and expression. In literature, AI models like GPT are transforming the landscape by assisting in the writing process, generating narrative ideas, dialogues, and even entire story arcs. These models can provide authors with creative prompts, help overcome writer's block, or offer new perspectives on narrative development, enriching the storytelling process. Exploring AI's role in creativity involves understanding how these technologies can enhance human creative processes while also examining the ethical and philosophical implications of AI-generated art. It raises questions about authorship, authenticity, and the essence of creativity. Future research should aim to refine AI's creative capabilities, ensuring that these technologies serve as catalysts for human creativity, fostering a symbiotic

relationship where AI and human ingenuity coalesce to create novel and previously unimaginable forms of artistic expression.

6. **Human-AI Collaboration**: Investigating ways to facilitate effective collaboration between humans and AI systems in various tasks. The exploration of human-AI collaboration delves into optimizing the synergy between human intelligence and artificial intelligence, ensuring that they complement each other to achieve outcomes that neither could independently. This area of research is crucial across numerous domains, from healthcare and design to decision-making processes in business and science, offering profound implications for enhancing productivity, innovation, and problem-solving. Effective human-AI collaboration hinges on developing AI systems that not only excel in task performance but also understand and adapt to human needs, preferences, and work styles. This involves AI systems being transparent in their operations and decision-making processes, allowing humans to comprehend, predict, and trust AI actions and suggestions. Moreover, these systems should be capable of interpreting human inputs accurately, whether they're verbal instructions, text-based commands, or physical interactions, to respond and adapt effectively. In domains like healthcare, collaboration could manifest as AI systems providing doctors with diagnostic assistance, treatment recommendations, or personalized patient care plans, always leaving the final judgment and empathetic interactions to human professionals. In creative fields, AI can offer artists and designers new ideas or assist in the iterative process of creation, while in business, AI could help in analyzing data to inform decision-making, identifying trends, or even predicting market shifts. Research should focus on developing intuitive interfaces for human-AI interaction, ensuring AI systems can learn from human feedback to refine their performance and alignment with human objectives. Additionally, exploring the psychological and social dimensions of human-AI interaction is vital, ensuring that these collaborations are not only productive but also enriching and satisfying for the human participants. By advancing our understanding and implementation of human-AI collaboration, we can unlock new horizons of collective intelligence, where the sum of human and artificial capabilities is greater than its parts, driving forward innovation and efficiency in myriad spheres of life.

7. **AI and Climate Change**: Studying how AI can contribute to climate modeling, environmental monitoring, and sustainable practices. AI's potential to address climate change is a crucial area of exploration, offering innovative solutions to one of the most pressing global challenges. By harnessing the power of AI, we can enhance climate modeling, improve environmental monitoring, and foster sustainable practices across industries, thereby contributing significantly to global efforts to mitigate and adapt to climate change. In climate modeling, AI can analyze complex datasets from diverse sources, including satellite imagery, sensor data, and

historical climate records, to predict future climate patterns with greater accuracy. These models can provide invaluable insights for policymakers, helping to forecast the impacts of various climate scenarios and inform more effective climate strategies and policies. Environmental monitoring benefits greatly from AI through the automation of data collection and analysis, enabling real-time tracking of environmental changes and potential hazards. AI can detect changes in ecosystems, monitor wildlife populations, or track the spread of pollutants, providing essential data that can lead to timely interventions and the protection of natural habitats. Moreover, AI can drive sustainable practices across various sectors. In energy, AI can optimize the distribution and consumption of renewable energy, improve efficiency, and reduce waste. In agriculture, AI-enabled precision farming can minimize resource use and maximize productivity, helping to reduce the sector's environmental footprint. In manufacturing, AI can streamline processes, reduce resource consumption, and promote circular economy principles, where waste is minimized, and materials are reused or recycled. Research in this domain should focus on developing and scaling AI solutions that are tailored to environmental challenges, ensuring they are accessible and implementable across different regions and contexts. Collaboration between AI researchers, environmental scientists, and policymakers is essential to align technological advancements with ecological goals, ensuring that AI's potential to combat climate change is fully realized, leading to a more sustainable and resilient future for our planet.

8. **AI in Finance**: Researching AI applications in financial markets, risk assessment, fraud detection, and algorithmic trading. AI's integration into the finance sector is profoundly reshaping its landscape, offering sophisticated tools for enhancing accuracy, efficiency, and predictive power in various financial operations. By delving into AI applications in financial markets, risk assessment, fraud detection, and algorithmic trading, researchers can uncover ways to not only optimize financial processes but also enhance security and compliance, providing a more robust financial ecosystem. In financial markets, AI systems can analyze vast arrays of market data, economic indicators, and news sources to forecast market trends, offering valuable insights for investors and financial institutions. By identifying patterns and correlations that might elude human analysts, AI contributes to more informed decision-making and strategic planning in investments. Risk assessment is another critical area where AI can make significant contributions. By processing complex datasets and historical information, AI models can predict the likelihood of defaults, bankruptcies, or other financial risks with greater precision. This capability allows banks and financial institutions to tailor their products and services more effectively, manage risks proactively, and comply with regulatory requirements. Fraud detection benefits immensely from AI's ability to rapidly analyze transaction data, identifying anomalies or patterns indicative of

fraudulent activity. By implementing AI-driven systems, financial entities can detect and prevent fraud more efficiently, safeguarding their operations and their customers' assets. Algorithmic trading is another frontier where AI's impact is substantial. AI algorithms can execute trades at optimal times based on analysis of market data, achieving speeds and efficiencies far beyond human capabilities. These algorithms can adapt to market changes in real-time, execute sophisticated trading strategies, and mitigate risks, contributing to more dynamic and responsive financial markets. Future research in AI's role in finance should focus on enhancing the accuracy and transparency of AI systems, ensuring they adhere to ethical standards and regulatory requirements. Addressing challenges related to data privacy, bias, and the potential for systemic risks is also crucial. By advancing AI applications in finance, the field can move towards more secure, efficient, and intelligent financial systems, benefiting economies and societies at large.

9. **AI in Legal and Ethical Decision-Making**: Examining the use of AI in legal processes, contract analysis, and ethical decision support. The incorporation of AI in legal and ethical decision-making represents a groundbreaking shift, offering new tools for legal professionals and enhancing the frameworks within which ethical decisions are made. AI's potential to transform legal processes, contract analysis, and ethical decision-making support is vast, promising to increase efficiency, accuracy, and access to justice. In legal processes, AI can automate and streamline routine tasks such as document review, legal research, and case prediction. By quickly analyzing large volumes of legal texts, AI can identify relevant precedents, suggest legal strategies, and even predict case outcomes, thereby assisting lawyers in preparing more effectively for cases. This not only reduces the time and cost associated with legal work but also allows legal professionals to focus on more nuanced and strategic aspects of their roles. Contract analysis is another area where AI is making significant inroads. AI systems can scrutinize contracts in detail, identifying potential issues, risks, and inconsistencies faster and more accurately than human review. This capability is invaluable in due diligence processes, mergers and acquisitions, and routine contract management, ensuring that legal agreements are sound, compliant, and aligned with clients' best interests. Beyond the practical applications in legal processes, AI holds promise for supporting ethical decision-making. By incorporating ethical frameworks into AI systems, they can assist in identifying potential ethical dilemmas, providing stakeholders with insights that might not be immediately apparent. This is particularly relevant in sectors like healthcare, business, and public policy, where complex ethical considerations are frequent. However, the integration of AI into legal and ethical domains also raises significant challenges, including questions about bias, transparency, and the potential displacement of traditional legal roles. Future research should therefore not only

focus on enhancing the capabilities and applications of AI in these fields but also on ensuring that these systems are developed and used in a manner that upholds the principles of justice, fairness, and ethical integrity. This includes rigorous testing for biases, clear accountability mechanisms, and ongoing dialogue among technologists, legal professionals, ethicists, and policymakers to ensure that AI serves to augment rather than undermine the legal and ethical foundations of society.

10. **AI and Robotics**: Investigating the integration of AI and GPT into robots and autonomous systems for various applications. The convergence of AI and robotics is forging a new frontier in technology, where robots and autonomous systems are not just tools but intelligent entities capable of learning, adapting, and interacting with their environment in unprecedented ways. The integration of AI, particularly GPT models, into robotics opens up a myriad of applications, ranging from industrial automation and healthcare to personal assistants and search-and-rescue operations. In industrial settings, AI-enhanced robots can perform complex tasks with high precision and flexibility, adapting to new processes or environments without extensive reprogramming. This adaptability, powered by AI's learning capabilities, allows for more efficient production lines, reduced downtime, and the ability to customize manufacturing processes on the fly. Healthcare is another domain where AI-driven robotics holds significant promise. Surgical robots, equipped with AI, can assist in complex procedures with precision beyond human capability. AI-driven diagnostic robots can navigate hospital environments independently, collecting and analyzing patient data to support healthcare providers in decision-making. Personal assistant robots, imbued with GPT's natural language processing abilities, can understand and respond to human language with nuance and context-awareness, providing companionship, assistance, and information to users in homes and workplaces. In the realm of search and rescue, AI-powered robots can navigate hazardous or inaccessible environments, making autonomous decisions to locate and assist individuals in distress, all while adapting to dynamic conditions and providing critical information to human responders. Future research in AI and robotics should focus on enhancing the symbiosis between AI's decision-making and learning capabilities and the physical dexterity and adaptability of robots. This includes improving robots' understanding of their environments, refining their interactions with humans and other machines, and ensuring their ethical and safe deployment. As AI and robotics continue to evolve, they hold the potential to transform every aspect of our lives, offering solutions to some of the world's most pressing challenges and enriching our daily experiences.

11. **AI and Language Translation**: Advancing machine translation models for better cross-language communication. The advancement of AI in language translation is pivotal for breaking down communication barri-

ers across the globe, fostering better cross-cultural understanding and cooperation. Machine translation models, especially those powered by AI technologies like GPT, are at the forefront of this transformation, striving to achieve translations that are not only accurate and fluent but also contextually and culturally nuanced. AI-driven translation models have made significant strides in understanding and translating between languages with high levels of accuracy. However, the quest for perfection continues as these models aim to master the subtleties of language, including idioms, slang, and regional dialects. The goal is to produce translations that are not just linguistically accurate but also culturally resonant, conveying the original message's intent, tone, and emotional nuance. Enhancing machine translation involves deep learning algorithms that can process vast datasets, learning from each translation to improve future performance. This iterative learning process allows AI models to understand language patterns, grammar, and vocabulary more deeply, refining their translation capabilities over time. Another critical area of research is real-time translation, which has profound implications for international communication, business, and diplomacy. Improving real-time translation accuracy and speed can facilitate more effective dialogue and collaboration across language divides, enabling more immediate and natural cross-lingual interactions. Future advancements in AI-driven translation will also focus on low-resource languages, which have traditionally been underrepresented in translation technology. Enhancing translation for these languages can preserve cultural heritage and provide global access to a broader range of knowledge and perspectives. In essence, the future of AI in language translation lies in creating more adaptive, intelligent, and inclusive translation tools. These advancements will not only enhance global communication but also contribute to a more interconnected and empathetic world, where language is no longer a barrier but a bridge connecting diverse communities and cultures.

12. **AI in Social Media Analysis**: Researching AI's role in social media sentiment analysis, content moderation, and recommendation systems. AI's integration into social media analysis represents a significant leap forward in understanding and navigating the vast, dynamic landscape of online interactions. By leveraging AI, platforms can analyze user sentiment, moderate content effectively, and enhance recommendation systems, thereby shaping more engaging, safe, and relevant online environments. Sentiment analysis through AI involves the nuanced interpretation of user-generated content to gauge public opinion, mood, or response to various topics, brands, or events. AI models, trained on vast datasets, can detect nuances in language, discerning not just the overt sentiment but also the underlying emotions and subtleties. This capability is invaluable for businesses, policymakers, and content creators aiming to understand and respond to audience feedback or market trends. Content moderation

is another critical area where AI is making a substantial impact. As social media platforms grapple with the sheer volume of user-generated content, AI tools are essential for identifying and addressing harmful or inappropriate content at scale. These systems can detect various types of problematic content, from hate speech and bullying to misinformation and explicit material, aiding human moderators, and ensuring a safer online environment. AI also plays a pivotal role in recommendation systems, which influence the content users encounter on social media. By analyzing user preferences, behavior, and interaction patterns, AI algorithms can curate personalized content feeds, enhancing user engagement and satisfaction. However, this also comes with the responsibility to avoid reinforcing biases or creating echo chambers, prompting ongoing research into more transparent and balanced recommendation algorithms. Future research in AI's application to social media analysis will likely focus on improving the accuracy and ethical considerations of these technologies. This includes enhancing the sensitivity of sentiment analysis, ensuring the fairness and accountability of content moderation, and developing recommendation systems that promote diversity and protect user privacy. As social media continues to evolve, AI will be central to fostering platforms that are not only more responsive and personalized but also committed to upholding the integrity and well-being of their digital communities.

13. **AI in Transportation**: Studying AI applications in autonomous vehicles, traffic management, and transportation optimization. AI's role in transforming transportation is profound, offering groundbreaking applications in autonomous vehicles, traffic management, and transportation optimization, which collectively promise to redefine mobility, enhance safety, and reduce environmental impact. In the realm of autonomous vehicles, AI is the linchpin technology that enables cars, trucks, and drones to navigate complex environments independently. AI systems process data from various sensors and cameras in real-time, making decisions about steering, speed, and route planning to ensure safe and efficient travel. The evolution of these systems not only promises to reduce human error—a leading cause of accidents—but also aims to optimize traffic flow, reduce congestion, and contribute to lower emissions. Traffic management is another crucial area where AI can have a significant impact. By analyzing data from a myriad of sources, including traffic cameras, sensors, and GPS data from vehicles, AI can predict traffic patterns, identify potential bottlenecks, and suggest optimal routing for traffic congestion mitigation. Such AI-driven systems can dynamically control traffic lights, adjust speed limits, and provide real-time updates to drivers, enhancing the overall efficiency and safety of road networks. Furthermore, AI plays a pivotal role in transportation optimization beyond roadways. It can enhance the scheduling and routing of public transportation, optimize logistics and supply chains, and even inform urban planning to create more efficient

and sustainable transportation ecosystems. Future research in AI and transportation will likely focus on improving the algorithms that drive autonomous vehicles, making them more reliable in diverse conditions and scenarios. It will also explore more integrated and intelligent traffic systems that can adapt to changing conditions and support the needs of growing urban populations. Additionally, the ethical and societal implications of autonomous transportation, such as job displacement, privacy concerns, and safety, will remain critical areas of investigation, ensuring that advancements in transportation AI align with broader societal values and priorities.

14. **AI in Business Process Automation**: Exploring how AI can streamline business operations, from customer support to supply chain management. AI's integration into business process automation is revolutionizing the way companies operate, offering smarter, more efficient approaches to a range of business functions, from customer support to supply chain management. This transformation not only boosts productivity and efficiency but also enables businesses to deliver enhanced, personalized services. In customer support, AI-driven chatbots and virtual assistants can handle a vast array of inquiries in real-time, providing instant responses to customer queries and freeing human agents to tackle more complex issues. These AI systems learn from each interaction, continuously improving their ability to resolve queries effectively and enhancing customer satisfaction. Supply chain management is another critical area benefiting from AI. By analyzing data from various touchpoints in the supply chain, AI can predict demand fluctuations, optimize inventory levels, and identify potential disruptions before they occur. This predictive capability allows businesses to make informed decisions, reduce waste, and maintain continuity in their operations, even in the face of unforeseen challenges. Moreover, AI can streamline administrative tasks such as data entry, invoice processing, and compliance checks. By automating these routine processes, companies can reduce the risk of human error, increase efficiency, and allow employees to focus on more strategic, value-adding activities. AI's role in business process automation also extends to financial analysis, where it can provide deep insights into financial health, identify cost-saving opportunities, and enhance decision-making. In marketing, AI can personalize customer experiences, optimize campaigns, and analyze consumer behavior to predict trends and tailor strategies accordingly. Future research in AI for business process automation will likely explore more sophisticated algorithms and models to enhance the adaptability and decision-making capabilities of AI systems. This includes ensuring that AI-driven automation is transparent, ethical, and aligns with the overall goals and values of the organization. As AI technologies evolve, they promise to unlock new levels of efficiency, agility, and innovation in business, driving competitive advantage and fostering sustainable growth.

15. **AI and Security**: Investigating AI's role in cybersecurity, threat detection, and anomaly detection. AI's integration into security, particularly in cybersecurity, threat detection, and anomaly detection, represents a pivotal advancement in safeguarding digital and physical environments. By leveraging AI's capabilities, organizations can proactively identify and respond to potential threats, enhancing their resilience against a wide array of security challenges. In cybersecurity, AI algorithms are instrumental in detecting patterns and anomalies that indicate potential threats or breaches. These systems can analyze vast quantities of data at an unprecedented speed, identifying suspicious activities that might elude human oversight. AI-driven cybersecurity can predict and neutralize threats, from phishing attempts to advanced persistent threats, ensuring data integrity and system security. Beyond the digital realm, AI plays a crucial role in physical security, particularly in threat detection and surveillance. AI-enabled cameras and sensors can monitor environments continuously, recognizing anomalous or dangerous behavior and triggering alerts more efficiently than traditional methods. This capability is invaluable in preventing criminal activities, enhancing public safety, and securing sensitive areas. Anomaly detection, a subset of AI's capabilities, extends across various domains, including finance, healthcare, and critical infrastructure. AI systems can identify outliers in data patterns that may signify fraud, equipment failures, or other risks, allowing for timely interventions to avert potential crises. Future research in AI and security will likely focus on enhancing the accuracy and predictive capabilities of AI systems while ensuring they respect privacy and ethical standards. As AI systems become more autonomous in their security functions, establishing trust and transparency in their operations is crucial. Additionally, as adversaries also adopt AI, the field must continuously innovate to stay ahead of increasingly sophisticated threats. By advancing AI's role in security, we can aspire to create a safer, more secure digital and physical world, where threats are identified and neutralized with unprecedented precision and speed.

A.2   TEACHING WITH AI

AI is powerful and will reshape the educational landscape in profound ways. Its influence spans a spectrum of applications, from redefining personalized learning experiences to streamlining the once time-consuming administrative tasks that educators grapple with daily. In this section, we examine the multifaceted roles AI plays within the dynamic sphere of education. By evaluating these AI-driven innovations, we not only gain insight into their transformative capabilities but also discern the profound implications they hold for educators and learners alike.

1. **AI as a teaching assistant**: Artificial Intelligence (AI) plays multifaceted roles in education, one of which is acting as a teaching assistant.

AI-powered tools and platforms can assist educators in various aspects of teaching, such as grading assignments, providing personalized feedback to students, and managing administrative tasks. For example, AI algorithms can analyze students' performance data to identify areas where they may need additional support or enrichment. Additionally, AI chatbots can answer students' questions in real-time, providing instant assistance and guidance. By automating routine tasks and providing personalized support, AI frees up educators' time, allowing them to focus on more meaningful interactions with their students.

2. **AI as a learning facilitator**: Beyond assisting educators, AI also serves as a learning facilitator, helping students acquire knowledge and develop essential skills. AI-powered tutoring systems can provide individualized learning experiences tailored to each student's unique learning style and pace. These systems can adapt to students' strengths and weaknesses, delivering customized learning materials and exercises to address their specific needs. Moreover, AI can create immersive learning environments through simulations, virtual reality, and interactive educational games. By engaging students in hands-on activities and experiential learning, AI fosters a deeper understanding and retention of academic concepts.

3. **AI as a content creator**: In addition to supporting teaching and learning processes, AI can also generate educational content, such as textbooks, lectures, and tutorials. Natural language generation (NLG) algorithms, like those used in GPT models, can produce high-quality written content on a wide range of topics. For instance, AI can generate educational materials tailored to students' grade levels, language proficiency, and learning objectives. Furthermore, AI can assist educators in developing multimedia presentations, interactive lessons, and digital learning resources. By leveraging AI-generated content, educators can access a wealth of educational materials to enhance their teaching practices and enrich students' learning experiences.

4. **AI as a learning assistant**: AI can serve as a personalized learning assistant, providing students with tailored feedback, explanations, and resources based on their individual learning needs and preferences. Through adaptive learning algorithms, AI platforms can track students' progress, identify areas for improvement, and offer targeted interventions to support their learning journey. For example, platforms like Khan Academy leverage AI algorithms to provide students with personalized practice exercises and instructional videos, helping them master concepts at the ir own pace.

5. **AI for collaborative learning**: AI tools can facilitate collaborative learning experiences by enabling students to work together on projects, engage in peer feedback, and participate in group discussions. These platforms leverage natural language processing capabilities to facilitate seamless

communication and collaboration among students, regardless of their physical location. For instance, tools like Google Workspace for Education provide collaborative document editing and real-time communication features, allowing students to collaborate on assignments and projects in virtual environments.

Recently, OpenAI has released some resources for teachers (*openai.com*). The following example shows applications of the one of these examples on building a comprehensive interactive tutor. Alternatively, users can choose GPT apps like Tutor Me and Math Solver.

EXAMPLE

The following prompt will turn ChatGPT (or GPT-4) into an interactive tutor:

Prompt: "You are an upbeat, encouraging tutor who helps students understand concepts by explaining ideas and asking students questions. Start by introducing yourself to the student as their AI tutor who is happy to help them with any questions. Only ask one question at a time.

First, ask them what they would like to learn about. Wait for the response. Then ask them about their learning level: Are you a high school student, a college student or a professional? Wait for their response. Then ask them what they know already about the topic they have chosen. Wait for a response.

Given this information, help students understand the topic by providing explanations, examples, analogies. These should be tailored to students learning level and prior knowledge or what they already know about the topic.

Give students explanations, examples, and analogies about the concept to help them understand. You should guide students in an open-ended way. Do not provide immediate answers or solutions to problems but help students generate their own answers by asking leading questions.

Ask students to explain their thinking. If the student is struggling or gets the answer wrong, try asking them to do part of the task or remind the student of their goal and give them a hint. If students improve, then praise them and show excitement. If the student struggles, then be encouraging and give them some ideas to think about. When pushing students for information, try to end your responses with a question so that students must keep generating ideas.

Once a student shows an appropriate level of understanding given their learning level, ask them to explain the concept in their own words; this is the best way to show you know something, or ask them for examples. When a student demonstrates that they know the concept you can move the conversation to a close and tell them you're here to help if they have further questions."

The following URL shows the result of execution of the abovementioned prompt, using GPT-4 for thermodynamics as an example:

https://chat.openai.com/share/fc289f8d-6972-448b-b36f-149e2ae389a8. Please note that trials may generate different responses.

ChatGPT:

"Hello there! ☺ I'm your AI-Tutor, and I'm thrilled to be here to assist you on your learning journey. Remember, there's no such thing as a silly question, and I'm here to help you every step of the way. Let's make learning fun and engaging!

To start, what topic or subject would you like to learn about today?

User: "Thermodynamics."

ChatGPT: "Fantastic choice! Thermodynamics is a fascinating field that delves into the principles governing energy and heat. Now, to tailor our discussion to your needs: Are you a high school student, a college student, or a professional?"

User: "A college student."

ChatGPT: "Great! As a college student, you probably have some foundational knowledge in physics and chemistry, which will be very helpful as we delve into thermodynamics. To get started, could you share with me what you already know about thermodynamics? This will help me understand your current knowledge level and guide our discussion accordingly."

User: "A first level college course on this topic."

ChatGPT: "Perfect! Since you've taken a first level college course in thermo-dynamics, you likely have an understanding of its basic principles and laws. Let's build on that foundation. Thermodynamics often starts with concepts like system and surroundings, types of systems (isolated, closed, and open), and moves on to the laws of thermodynamics.

Could you tell me which concept or area within thermodynamics you find intriguing or perhaps challenging? We can focus on that to enhance your understanding."

User: "Second law, with clear examples to help the concept."

ChatGPT: "Absolutely! The second law of thermodynamics is a key principle, and it's all about the direction of processes and the concept of entropy. In essence, the second law states that the total entropy of an isolated system can never decrease over time. It can only increase or remain constant. This law gives us a direction for the flow of energy and the progression of processes in the universe. A classic example to illustrate the second law is the melting of ice in a room-temperature environment. When you take ice (a more ordered, low-entropy state) and place it in a room (a warmer, higher-entropy environment), the ice absorbs heat and melts into water, spreading out in a more disordered state, increasing the system's total entropy.

Here is a question to ponder: Why doesn't the water spontaneously freeze back into ice in the room, even though the water molecules could theoretically arrange themselves back into a solid state? What do you think this implies about the second law of thermodynamics?"

Here, we stop the conversation. Users may want to continue this conversation or challenge the tutor with another topic.

Your Personal Interactive Tutor

Users may want to build a GPT app using the comprehensive prompt as found in the Configuration. We created a similar app named "Your Personal Interactive Tutor." This app is available at *https://chat.openai.com/g/g-6iMl BOH0g-your-personal-interactive-tutor.* Users must have access to GPT-4 to use this app.

A.3   MULTIMODAL GPT-4 WITH VOICE AND IMAGE AS PROMPTS

With the advent of GPT-4, the capabilities of AI have expanded significantly, since it introduced sophisticated "see, hear, talk" abilities that mark a new era of multimodal interaction. The "see" feature refers to GPT-4's ability to interpret and understand visual inputs through images, allowing it to analyze and discuss content presented in a picture or diagram, enhancing its utility in fields such as medical imaging and art critique. The "hear" aspect encompasses the model's proficiency in audio processing, enabling it to understand spoken language, recognize different sounds, and even interpret emotional subtleties in speech, thereby broadening its applications to include real-time transcription, language learning, and emotional analysis. Finally, the "talk" feature signifies GPT-4's advanced natural language generation capabilities, producing text that is not only contextually relevant and syntactically correct but also nuanced and expressive in a way that closely mimics human speech. Together, these "see, hear, talk" features transform GPT-4 into a more holistic, interactive, and adaptable AI model, capable of engaging in a wide range of tasks that require a complex understanding of the world and human communication.

Recently, OpenAI has introduced Sora (video generation models as world simulators, *openai.com*), an AI model that can create realistic and imaginative video scenes from text instructions.

Recently, OpenAI introduced *Sora*, a text-to-video model. Sora can generate videos up to a minute long while maintaining visual quality and adherence to the user's prompt. This AI model can create realistic and imaginative scenes from text instructions.

TopAI.Tools

TopAI.Tools (*https://topai.tools/*) is a leading AI tools directory and search engine. It allows users to discover the AI tools, apps, and services to elevate their businesses. The platform provides a comprehensive list of AI tools and resources, making it easier for users to find the right solutions for their specific needs.

A.4 COMMON PITFALLS AND SOLUTIONS

During the process of innovation, there inevitably arise missteps, oversights, and unforeseen challenges. These common pitfalls, though often seen as deterrents, can be transformed into opportunities for growth and learning when approached with the right solutions. This section discusses the most frequent problems encountered in various endeavors and provides practical solutions to overcome them. By acknowledging these challenges and equipping ourselves with the tools to navigate them, we can make more informed decisions, more resilient systems, and a brighter future shaped by both our successes and our failures.

Here is a list (not exclusive) of common pitfalls in the context of prompt engineering for such models:

1. **Ambiguity:** A lack of specificity in prompts can lead to generalized or off-target responses from the AI.

2. **Overloading:** Providing too much information or too many requests in a single prompt can confuse the model and dilute the output's relevance.

3. **Assuming Prior Context:** Unlike human conversations, if not specifically designed for it, models like ChatGPT do not maintain context across separate prompts. Assuming they remember past interactions is a mistake.

4. **Over-Reliance on Model Accuracy:** While GPT models are powerful, they can make factual or logical errors. Blindly trusting outputs without verification can lead to being misinformed.

5. **Lack of Clarity in Instructions:** Not being explicit about the desired format or depth of the response can lead to misaligned outputs.

6. **Not Anticipating Variability:** Similar prompts can yield slightly different answers on different invocations. Not accounting for this variability can be problematic in applications requiring consistent responses.

7. **Ignoring Potential Biases:** All models have biases based on their training data. Not being aware or not checking for these biases in responses can perpetuate misinformation or stereotypes.

8. **Overcomplication:** Crafting prompts that are too complex or intricate can sometimes backfire, leading the model to produce convoluted or irrelevant responses.

9. **Neglecting Iterative Refinement:** Prompt engineering often requires iterative refinement. Sticking rigidly to an initial prompt without tweaking it based on outputs can reduce its effectiveness.

10. **Overestimating Model Understanding:** While GPT models can generate human-like text, they do not truly "understand" content. Expecting human-level comprehension or intuition can lead to overestimating the model's capabilities.

Being aware of these problems is crucial for anyone aiming to harness the full potential of ChatGPT and similar models effectively. Proper prompt engineering can mitigate many of these challenges, leading to more accurate and relevant AI-generated content.

GLOSSARY

1. **Prompt:** A specific instruction or input provided to an AI model to generate a desired response or output.

2. **AI (Artificial Intelligence):** The simulation of human intelligence processes by machines, including learning, reasoning, problem-solving, and decision-making.

3. **LLM (Large Language Model):** Refers to AI models like GPT-4 that have a massive number of parameters, enabling them to perform advanced natural language understanding and generation tasks.

4. **Transformer Architecture:** A neural network architecture widely used in natural language processing tasks, known for its ability to handle sequential data efficiently.

5. **Chatbot:** An AI-driven program designed to engage in text-based or voice-based conversations with users, often for customer support or information retrieval.

6. **Contextual Prompt:** A type of prompt that includes relevant information or context to guide AI responses, enhancing the model's understanding.

7. **Multimodal Prompt:** A prompt that combines multiple forms of input, such as text, images, or audio, to generate more diverse and context-aware responses.

8. **Bias and Fairness:** Concerns related to AI models producing responses that are discriminatory or exhibit biases based on race, gender, or other characteristics.

9. **Plugin:** An extension or module that can be added to an AI model to provide additional functionalities or domain-specific capabilities.

10. **Ethical AI:** The practice of developing and using AI systems in a manner that adheres to ethical principles, respects human rights, and promotes fairness and transparency.

11. **Human-AI Collaboration:** The interaction and cooperation between humans and AI systems to achieve specific tasks or goals.

12. **Prompt Engineering:** The process of crafting effective prompts to guide AI model behavior and generate desired responses.

13. **Conditional Logic:** Techniques used to condition AI responses based on specific input criteria or constraints.

14. **Creative Writing Prompts:** Prompts designed to inspire creative content generation, such as stories, poetry, or artwork.

15. **Meta-Prompt:** A high-level prompt that guides the generation of other prompts, often used for task automation or orchestration.

16. **Natural Language Processing (NLP):** A field of AI and linguistics focused on the interaction between computers and human language, including text analysis and generation.

17. **Feedback Loop:** The iterative process of providing feedback to AI models to improve their performance and responses.

18. **Human-in-the-Loop:** An AI system that involves human oversight, intervention, or decision-making in combination with automated AI processes.

19. **Responsible AI:** The practice of developing AI systems that are accountable, transparent, and aligned with ethical and societal values.

20. **AI Safety:** Efforts and techniques aimed at ensuring the safe and responsible deployment of AI systems.

21. **OpenAI:** A research organization and company known for developing state-of-the-art AI models, including the GPT series.

22. **End-to-End Learning:** A machine learning approach where a model learns directly from raw input to produce desired output without manual feature engineering.

23. **Transfer Learning:** A technique where a pre-trained model is adapted for a specific task, reducing the need for extensive training data.

24. **Fine-Tuning:** The process of adjusting a pre-trained model's parameters on a task-specific dataset to improve performance.

25. **Neural Network:** A computational model inspired by the structure and function of the human brain, used in various AI applications.

26. **Deep Learning:** A subset of machine learning that utilizes neural networks with many layers (deep architectures) for complex tasks.

27. **Reinforcement Learning:** A machine learning paradigm where agents learn to make decisions by interacting with an environment and receiving rewards or penalties.

28. **Semantic Analysis:** The process of extracting meaning and context from text, often used in NLP tasks.

29. **Model Architecture:** The structural design of a neural network, defining the arrangement and connections of its layers.

30. **Overfitting:** When a machine learning model performs well on training data but poorly on new, unseen data due to excessive complexity.

31. **Token:** A single unit of text obtained through tokenization. Tokens can be words, sub-words, or characters, depending on the tokenization method used.

32. **Tokenization:** The process of breaking down text into smaller units called tokens, which can be words, phrases, or characters. Tokenization is a fundamental step in natural language processing and allows for the analysis of text at a granular level.

REFERENCES

1. T. B. Brown et al., "Language Models are Few-Shot Learners," in NeurIPS, 2020.

2. Radford, J. Wu, R. Child, D. Luan, D. Amodei, and I. Sutskever, "Language Models are Unsupervised Multitask Learners," OpenAI Blog, [Online]. Available: *https://cdn.openai.com/better-language-models/language_models_are_unsupervised_multitask_learners.pdf.* [Accessed: June 15, 2023].

3. T. B. Brown, B. Mann, N. Ryder, M. Subbiah, J. Kaplan, P. Dhariwal, ... and D. Amodei, "Language models are few-shot learners," in *arXiv preprint arXiv:2005.14165*, 2020. [Online]. Available: *https://arxiv.org/abs/2005.14165.* (Accessed: Aug. 27, 2023).

4. D. Adiwardana et al., "Towards a Human-like Open-Domain Chatbot," arXiv:2009.03300, 2020.

5. J. Dodge et al., "Fine-tuning Pre-trained Language Models: Weight Initializations, Data Orders, and Early Stopping," arXiv:2002.05202, 2020.

6. S. Roller et al., "Recipes for Building an Open-Domain Chatbot," in arXiv preprint arXiv:2002.07033, 2020.

7. A. Holtzman et al., "The Curious Case of Neural Text Degeneration," arXiv:1904.09751, 2019.

8. K. VanLehn, "The relative effectiveness of human tutoring, intelligent tutoring systems, and other tutoring systems," Educational Psychologist, vol. 46, no. 4, pp. 197-221, 2011.

9. J. R. Anderson and C. D. Schunn, "Implications of the ACT-R learning theory: No magic bullets," in *Advances in Instructional Psychology*, R. Glaser, Ed. Lawrence Erlbaum Associates, 2000, pp. 1-34.

10. I. Roll, V. Aleven, B. M. McLaren, and K. R. Koedinger, "Improving students' help-seeking skills using metacognitive feedback in an intelligent tutoring system," *Learning and Instruction*, vol. 21, no. 2, pp. 267-280, 2011.

11. P. C. Verhoef, K. N. Lemon, A. Parasuraman, A. Roggeveen, M. Tsiros, and L. A. Schlesinger, "Customer experience creation: Determinants, dynamics and management strategies," in *Journal of Retailing*, vol. 85, no. 1, pp. 31-41, 2009.

12. T. H. Davenport, J. Harris, and J. Shapiro, "Competing on analytics: The new science of winning," Harvard Business Press, 2010.

13. Y. Zhang and P. Daugherty, "AI and machine learning in business: How it is changing the world," *MIS Quarterly Executive*, vol. 17, no. 2, 2018.

14. GPT-3 Powered Apps and Use Cases. ChatGPT. *https://chat.openai.com/*

15. Y. Liu, et al., "Multimodal Few-Shot Learning with Frozen Language Models," arXiv:2105.08050, 2021. [Online]. Available: *https://arxiv.org/abs/2105.08050*.

16. OpenAI. (n.d.). Prompt Engineering Playground. *https://platform.openai.com/playground/prompt-engineering*

17. Google Cloud. (n.d.). Dialogflow. *https://cloud.google.com/dialogflow*

18. Google Translate. (n.d.). *https://translate.google.com/*

19. Copy.ai. (n.d.). *https://www.copy.ai/*

20. E. M. Bender, T. Gebru, A. McMillan-Major, and S. Shmitchell, "On the Dangers of Stochastic Parrots: Can Language Models Be Too Big?" [Online]. Available: *https://faculty.washington.edu/ebender/papers/Stochastic_Parrots.pdf*, 2021.

21. K. Crawford and R. Calo, "There is a blind spot in AI research," *Nature*, vol. 538, no. 7625, pp. 311-313, 2016.

22. R. Tatman, "Gender and Dialect Bias in YouTube's Automatic Captions," in *Proceedings of the First ACL Workshop on Ethics in NLP*, 2017, pp. 98-107.

23. L. Floridi et al., "AI4People—an Ethical Framework for a Good AI Society: Opportunities, Risks, Principles, and Recommendations," *Minds and Machines*, vol. 28, no. 4, pp. 689-707, 2018.

24. A. Jobin, M. Ienca, E. Vayena, and others, "The global landscape of AI ethics guidelines," *Nature Machine Intelligence*, vol. 1, no. 9, pp. 389-399, 2019.

25. N. Diakopoulos, S. A. Friedler, M. Arenas, S. Barocas, M. Hardt, and A. Narayanan, "Principles for Accountable Algorithms and a Social Impact Statement for Algorithms," in Proceedings of the AAAI Conference on Artificial Intelligence, 2019, pp. 3302-3309. DOI: 10.1609/aaai.v33i01.33023309.

26. E. R. Mollick and L. Mollick, "Practical AI for Teachers and Students," Aug. 4, 2023. [Online]. Available: *https://www.youtube.com/playlist?list=PLwRdpYzPkkn302_rL5RrXvQE8j0jLP02j*.

27. E. R. Mollick and L. Mollick, "Assigning AI: Seven Approaches for Students, with Prompts," June 12, 2023. [Online]. Available: *http://dx.doi.org/10.2139/ssrn.4475995*.

28. E. R. Mollick and L. Mollick, "Using AI to Implement Effective Teaching Strategies in Classrooms: Five Strategies, Including Prompts," March 17, 2023. [Online]. Available: *http://dx.doi.org/10.2139/ssrn.4391243*.

29. 12 Resources to Master Prompt Engineering, Published on May 8, 2023 In *Mystery Vault*, accessed Dec. 2023, *https://analyticsindiamag.com/12-resources-to-master-prompt-engineering/*

30. AI: Grappling with a New Kind of Intelligence, *World Science Festival*, accessed Dec. 2023, *https://youtu.be/EGDG3hgPNp8*

INDEX

www.ingramcontent.com/pod-product-compliance
Lightning Source LLC
LaVergne TN
LVHW022322060326
832902LV00020B/3621